An Introduction to SSADM Version 4

THE McGRAW-HILL INTERNATIONAL SERIES IN SOFTWARE ENGINEERING

Consulting Editor

Professor D. Ince
The Open University

Titles in this Series

An Introduction to SSADM Version 4

Caroline Ashworth and Laurence Slater
of AIMS Systems Ltd

McGRAW-HILL BOOK COMPANY

London · New York · St Louis · San Francisco · Auckland · Bogotá
Caracas · Hamburg · Lisbon · Madrid · Mexico · Milan · Montreal
New Delhi · Panama · Paris · San Juan · São Paulo · Singapore
Sydney · Tokyo · Toronto

Published by
McGRAW-HILL Book Company Europe
Shoppenhangers Road · Maidenhead · Berkshire · SL6 2QL England
Telephone (0628) 23432
Fax (0628) 770224

British Library Cataloguing in Publication Data

Ashworth, Caroline
Introduction to SSADM Version 4. –
(International Software Engineering
Series)
I. Title   II. Slater, Laurence   III. Series
658.40380285421

ISBN 0–07–707725–3

Library of Congress Cataloging-in-Publication Data
Ashworth, Caroline
Introduction to SSADM version 4 / Caroline Ashworth and
Laurence Slater.
        p.   cm. – (The McGraw Hill international series in software
engineering)
Includes index.
ISBN 0–07–707725–3
1. Electronic data processing–Structured techniques.   2. System
analysis.   3. System design.   I. Slater, Laurence.
II. Title.   III. Series.
QA76.9.S84A85   1992
004.2′1–dc20   92–46102
                                                                              CIP

1234   9543

Typeset by TecSet Ltd, Wallington, Surrey
and printed and bound in England by Clays Ltd, St Ives plc

10.10.94

Contents

Preface

SSADM (Structured Systems Analysis and Design Method) has become, since its launch in 1981, the most widely used method for analysis and design in the United Kingdom. Originally used on projects within the UK government sector, it is now also gaining popularity within UK private industry. As one of the leading European methods, it is one of the principal inputs to the development of Euromethod, the initiative to provide a framework for analysis and design methods within Europe.

Since its inception SSADM has gone through a number of versions, each of which has further extended and enhanced the method. Version 4, the latest one, was released in May 1990 after an extensive period of development. This work was sponsored by the Government Centre for Information Systems (CCTA), much of the work being undertaken by a consortium of companies including AIMS Systems, Softlab and BT. We both played a leading role in this consortium, working for nearly a year on the Version 4 development. The CCTA were supported throughout the development by consultancy support from Model Systems, principally in the person of Phil Lomax.

The supporting infrastructure of SSADM consists of a number of bodies, publications and schemes. There is a Design Authority Board who approve all changes and improvements to the method, a British Standard for SSADM Version 4 is under development which will assist organizations with the procurement of SSADM services (consultancy, training, etc.), an SSADM Users' Group exists to bring together individuals and organizations interested in various aspects of the method, a Tools Conformance Scheme should be approved in 1993, and there is a growing number of publications published by individuals and the CCTA to supplement the SSADM Reference Manuals.

This book aims to give an overview of SSADM Version 4, to summarize the main 'facts' of the method, concentrating on the 'what' rather than the 'how'. It does so by concentrating on describing each of the techniques used in the method in turn. This description is preceded by a general introduction to the method and a description of the structure of SSADM and is followed by a series of appendices. Each of the technique chapters follows the same structure:

- *Introduction*, which gives an overview of the technique.
- *Naming conventions*, which introduces the terms used within the technique and acts as a glossary for the technique.
- *Place in structure*, which shows the position of the technique within the SSADM structure.
- *Place in Product Breakdown Structure*, which shows how the products of the techniques fit into the overall structure of the products of SSADM.

- *Notation and use*, which is the major part of the chapter and is a full description of the technique.
- *Comparison with Version 3*, which relates the technique back to its equivalent in Version 3 and shows how it has changed within Version 4.

Each of the techniques is demonstrated by use of a fictional case study known as the Opera Booking System. A general introduction to this case study can be found in Appendix F. The techniques with little or no change from previous versions of the method have been described in less detail than those that are new to Version 4, but we have aimed to include all the main points about each technique, nevertheless. As such, we would recommend this book as a useful revision text for the SSADM Version 4 Certificate of Proficiency examinations or as a quick reference for practitioners.

Acknowledgements and thanks are due to a number of individuals and organizations who have helped in our development of this book. Our employers, AIMS Systems, provided encouragement, facilities and, most importantly, much of the time needed to write this book. Among the individuals who provided assistance a very special thanks must go to Mark O'Brien who provided comprehensive comments and helpful suggestions throughout. Thanks also go to Peter Elliott, David Hitchings, Dan Jefferis and Chris Blackwell for a number of useful comments and assistance.

Note
This publication has been assessed by the Technical Committee of the SSADM Users Group as conformant to core SSADM Version 4 and this has been endorsed by the Design Authority Board.

1. Introduction

1.1 Background

The Structured Systems Analysis and Design Method (SSADM) has been in use since 1981 and has gained a standing as the most widely used method for systems analysis and design in the United Kingdom. Its development has been controlled to ensure that the method has kept in step with 'best practice' on projects, as well as taking into account developments in information systems technology. It has been used successfully on projects of all sizes in a variety of environments, and although its roots are in UK government departmental computer projects, it has been adopted by a growing number of public and private sector organizations worldwide.

1.2 Purpose of SSADM

SSADM was developed as the standard method for the analysis and design of information systems in UK Government departments. It was commissioned and promulgated by the CCTA (the Government Centre for Information Systems) who initially offered practical consultancy support to all government projects using SSADM. The CCTA had specific objectives that SSADM should satisfy, including:

- improved project planning and control;
- more effective use of experienced and inexperienced staff;
- better-quality systems;
- resilience to loss of key staff;
- supportable by computer-based tools;
- good communication between developers and end-users.

These objectives have been realized in a growing number of projects as SSADM has been put into practice and developed through a number of versions.

Each version of SSADM has successfully achieved these objectives to an increasing extent, and in SSADM Version 4 features have been introduced to ensure that some are achieved more effectively than previous versions.

Improved project planning and control

Splitting a project down into modules, stages, steps, and tasks gives project managers a clear picture of how an SSADM project should proceed and what should be done in order to achieve milestones in a project. Therefore, it is possible to plan on the basis of activities if this is the chosen method of planning.

Each module of SSADM Version 4 has a stated objective and a clearly defined set of products. This supports planning based upon products. It means that planning of specific tasks can be flexible, provided that the objectives are met and the products delivered to a predefined standard of quality.

SSADM Version 4 has been designed to provide the information required by the PRINCE project management method in its production of a Product Breakdown Structure, Product Descriptions, and a Product Flow Diagram. It is, however, comprehensive and flexible enough to fit into any style or standard of project management.

Effective use of experienced and inexperienced staff

SSADM incorporates a number of techniques into a clearly defined framework. This means that what needs to be done is very clearly defined. Although the successful implementation of SSADM still requires skilled analysts and designers, there will always be tasks that can be delegated to more inexperienced staff as the results of their work will be visible and checkable against other work done.

The introduction of additional diagrammatic techniques in Version 4 for the specification and design of processing means that the products developed are even more easy to check and integrate with others than their predecessors.

Better-quality systems

'Quality' of a system can be defined as 'fitness for purpose' in that it:

- satisfies the users' requirements;
- is delivered on time;
- is developed within budget.

The use of SSADM can contribute to the quality of a system by producing documentation at all stages of development that can be checked by the users to ensure that their requirements are being understood, interpreted and implemented in the correct way. The inclusion of Specification Prototyping techniques during the Definition of Requirements gives additional confidence that the users understand what is being proposed and that the developers understand what the users require.

The modules, stages, and steps of SSADM give a framework within which quality assurance can be applied. The Dictionary introduced into SSADM Version 4 provides quality criteria for all SSADM products that can be used in their quality control. These should be regarded as a starting point that must be reviewed and modified for each project.

Resilience to loss of key staff

The use of SSADM ensures that a large proportion of the information collected and modelled during the project is documented in a way that can be understood by staff subsequently assigned to a project. This should ensure that the loss of key staff will not mean that the project has to be restarted from scratch. The widespread use of SSADM also means that staff with SSADM skills can normally be recruited in the event of staff being lost and additional investment in training and the 'learning curve' should be reduced.

Supportable by computer-based tools

SSADM is well defined in terms of the techniques employed and the products developed. The techniques are defined in terms of notation standards and syntax rules and can be readily supported by computer-based tools. Many of the techniques produce graphical representations of the information that are especially well suited to computer-based support. SSADM products are defined in terms of their information content and cross-references to other products and these aspects can be built into a computer-based tool.

The progression through the different versions of SSADM has been towards tighter definitions and clearer description of the 'rules' of the method. Version 3 was expressed in terms of an entity–relationship model of concepts to assist tools developers and Version 4 products are being closely defined in a British Standard. The tools conformance scheme first introduced for SSADM Version 3 gives rules that the tools must conform to and evaluates tools to show the degree of support they can offer to SSADM. The conformance scheme for Version 4 will be based upon the definition in the British Standard.

There are a growing number of tools designed to support SSADM and the clearer definitions of Version 4 can only assist in the continuing development of automated support for SSADM.

Good communication between developers and end-users

There is often a great difference in the way users and developers view a computer system. Both groups use jargon the other finds unfamiliar and communication can often prove difficult. It is very important, therefore, that each group is able to check their understanding of what the other has said so that they are not misinterpreted. This communication is supported by the use of diagrammatic techniques that model and interpret the information passed between user and developer and express it in a way that is readily checked.

The SSADM techniques and procedures help communication in two ways:

- They prompt developers to ask the right questions and to make sure they have a comprehensive understanding of what is required.
- They provide the user with a means of checking the interpretation of what they have stated as their requirements.

1.3 Overview of SSADM Version 4

SSADM and the system development lifecycle

SSADM is used in the development of systems but it does not cover the whole system lifecycle. Its starting point is assumed to be after the completion of a strategy or scoping study and an initial 'setting up' of the project and its end point is the production of a detailed physical design of the system. Its coverage of the stages of the development lifecycle are represented in Fig. 1.1.

SSADM is often used in conjunction with other methods such as strategic planning, structured programming, and structured testing methods in order to cover the complete lifecycle. In addition, a project management method is generally applied to control and monitor the whole development lifecycle. Thus, SSADM is not a complete 'project'

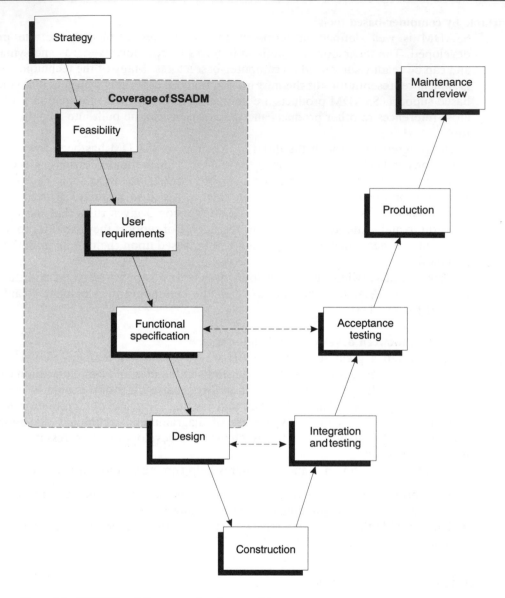

Figure 1.1 SSADM and the system development lifecycle

method even in the part of the lifecycle it covers. It needs to be supplemented by other techniques and procedures that cover aspects of a project other than the analysis and design tasks. These are referred to as Project Procedures within SSADM and are described briefly to show where they influence, and are influenced by, the techniques and procedures of SSADM.

SSADM Version 4 elements

SSADM specifies a number of techniques and procedures that fit into a framework of modules, stages, and steps. The techniques and procedures are used to produce a number of products at specific points in the framework. Interfacing with SSADM are a number of activities that are not strictly covered by SSADM but are of direct relevance to the conduct of an SSADM project.

SSADM Version 4 is described by four major components:

- The *structural model*, which describes the framework of modules, stages, and steps, together with inputs, outputs, and influences.
- *Techniques and procedures*, which define how the steps are to be performed.
- The *dictionary*, which defines all the products of SSADM together with associated quality criteria.
- *Project procedures*, which are the non-SSADM activities that are performed in the same part of the system development lifecycle and are relevant to the practice of SSADM.

These components are described in more detail in the subsequent chapters of this book and are described in overview below.

THE STRUCTURAL MODEL

The structure of SSADM Version 4 is based on modules, each of which covers one or two stages. Each stage is then further broken down into a series of steps. The full structural model can be found in Appendix A and is described in Chapter 2.

The five modules of SSADM Version 4 and their constituent stages are shown in Fig. 1.2. The structure of the method illustrates the approach of SSADM to the analysis and design of systems:

1. The requirements for the new system are analysed by first studying the current environment to gain an understanding of the business constraints and organization within which the system will be required to operate.
2. This understanding of the environment is used, together with the statement of requirements given by users, to build an outline picture of the new system before specifying anything in detail.
3. The outline description of the new system is used as a basis for a detailed examination of the requirements, which involves modelling various different aspects of the system. These models are cross-checked with one another and some of the functions may be prototyped to gain a clear picture of what is needed.
4. A detailed logical design of the processing and dialogue aspects of the system are then developed, in parallel with the final selection of the implementation platform for the system.
5. The complete logical specification for the system is finally converted into a design that is specific to the hardware/software configuration chosen for the system.

There is no explicit dependency between modules even though, in practice, they will be performed in a defined sequence. If a module consists of more than one stage, there may be a direct dependency between the stages, or they may be performed in parallel.

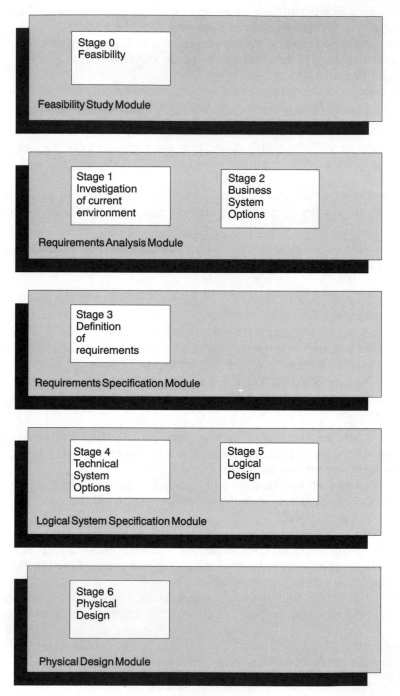

Figure 1.2   Modules and stages of SSADM Version 4

The steps of SSADM are described in the reference manuals in terms of their:

- objectives;
- participants;
- activities/tasks to be performed;
- preconditions (inputs and references);
- products;
- techniques used.

The SSADM structural model is very comprehensive and can be tailored to specific project circumstances. One of the first activities of the method is to develop a structural model specific to the project that uses only those parts of the standard model that will give benefits to the project. This is carried out by taking into account the type of development, the type of application, and the specific starting point of the project. For example, if there is no current system or the project is a small enhancement to a current system, the full set of procedures will not be appropriate.

TECHNIQUES AND PROCEDURES

The techniques and procedures of SSADM describe how aspects of the system can be modelled and how information should be collected and interrelated. The techniques that are diagrammatic are described in terms of their syntax and notation as well as their application at specific points in the method. The non-diagrammatic techniques are described in terms of procedures that should be followed.

The diagrammatic techniques of SSADM Version 4 are as follows:

- Logical Data Modelling;
- Data Flow Modelling;
- Entity/Event Modelling (Entity Life History Analysis and Effect Correspondence Diagramming);
- Enquiry Access Paths;
- I/O Structuring;
- Dialogue Design (Dialogue Structures and Menu Structures);
- Logical Database Process Design.

The techniques and procedures that are non-diagrammatic include:

- Relational Data Analysis;
- Requirements Definition;
- Function Definition;
- Formulation of Options (some diagrammatic techniques may be applied here);
- Specification Prototyping.

In addition, guidelines are given for Physical Design which involve some diagrammatic techniques for database design but which are principally general in character.

The naming conventions, context, notation, and use of each of these techniques is described in Chapters 3–13 of this book.

THE DICTIONARY AND PRODUCT BREAKDOWN STRUCTURE

The SSADM Version 4 dictionary describes all the SSADM products, both diagrammatic and non-diagrammatic, in terms of the standards to which they should be produced, their information content, and their interrelationships. The complete list of products from the dictionary is given in Appendix C.

The Product Breakdown Structure indicates the groupings of products that are deliverable from each module. The supporting documentation for a diagram, for example, is always grouped together with the diagram in the Product Breakdown Structure. Each of the chapters describing the techniques and procedures includes the part(s) of the Product Breakdown Structure relevant to that procedure or technique. The complete Product Breakdown Structure for SSADM Version 4 can be found in Appendix B.

PROJECT PROCEDURES

The project activities that are complementary to the SSADM techniques and procedures are termed 'project procedures'. They are described briefly within SSADM in order to:

- describe their influences upon the practice of SSADM;
- show where in the SSADM structure they are of relevance;
- understand the context of SSADM within the project as a whole.

The project procedures described within SSADM are:

- project management;
- quality assurance;
- risk assessment and management;
- capacity planning;
- testing;
- training;
- take-on;
- (technical) authoring;
- standards.

The formal mechanism for communication between SSADM and project procedures is the information highway. The project procedures and information highway are described in more detail in Chapter 14.

1.4 Comparison with Version 3

SSADM Version 4 represents a major change from SSADM Version 3. Readers familiar with Version 3 may find it difficult to work out exactly what has changed and what is still familiar from Version 3. In many cases, changes are relatively cosmetic in terminology and notation but the basic use of a technique may be very similar to previous versions. In other cases, the terminology may be similar to that used in previous versions but the approach and substance may have been changed completely.

To help those of you who are familiar with Version 3, we have included a brief section at the end of most chapters giving an outline comparison of the Version 4 concepts and their equivalent in Version 3. Hopefully, this will give you an appreciation of where the major changes are and where you are on familiar ground!

Appendix E gives a direct comparison between Version 4 techniques and their Version 3 equivalents.

2. The Structure of SSADM Version 4

2.1 Introduction

The structure of SSADM Version 4 is based on the concept of *modules*. Each module has a defined purpose and a set of products that are detailed in the Product Breakdown Structure. There is no explicit dependency between modules even though, in practice, they will be performed in a defined sequence. The modules of SSADM Version 4 are shown in Fig. 2.1. This shows that there are no direct links between modules but that products from one module are used as an input to the next module.

The reason for defining modules independently of one another is to make the method tailorable to specific project needs. Some types of project may not need to use all of the modules or may find it preferable to replace whole modules with alternative procedures more suited to their environment. The discrete nature of the SSADM modules with clearly defined objectives and products makes this possible. Also a project can choose to undertake, say, only the Requirements Analysis Module to ascertain the possible requirements for a system without necessarily undertaking the further modules.

Each module encompasses either one or two *stages*, as shown in Table 2.1. Each stage is then broken down into a number of constituent *steps*. The steps within each stage and their interdependency are detailed in Appendix A.

The complete set of modules, stages, and steps, together with their descriptions, is called the *structural model*.

2.2 Feasibility Study Module

The purpose of the Feasibility Study Module is to assess whether or not a particular project should proceed before any major planning or expenditure is undertaken and to define roughly the boundaries and objectives of the proposed system.

Overview

A feasibility study may be optionally undertaken as a precursor to the full analysis and design of a new system. The study describes in overview the functionality and data requirements of the new system together with broad technical and business options for the system. This helps to establish the viability of the system in terms of a number of factors, including:

- Will the system meet business objectives?
- Do the benefits outweigh the costs?
- What size of system will be required?

9

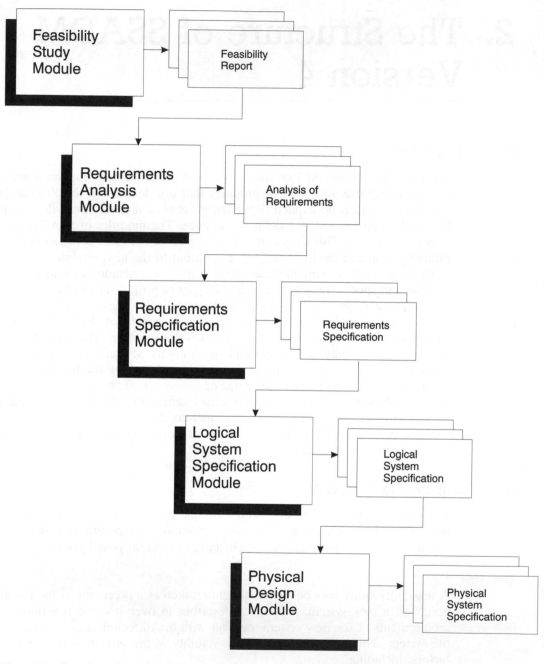

Figure 2.1 Overview of modules and products

Table 2.1

Module	Stage
Feasibility	0 Feasibility
Requirements Analysis	1 Investigation of Current Environment
	2 Business System Options
Requirements Specification	3 Definition of Requirements
Logical System Specification	4 Technical System Options
	5 Logical Design
Physical Design	6 Physical Design

As well as employing the techniques of SSADM, the conduct of a feasibility study will draw on skills and techniques that are not defined within SSADM and will require the application of business analysis experience.

Figure 2.2 gives an overview of the steps within Stage 0: Feasibility. It should be noted that the Feasibility Study Module is not described further in the subsequent chapters of this book.

Products

Feasibility Report.

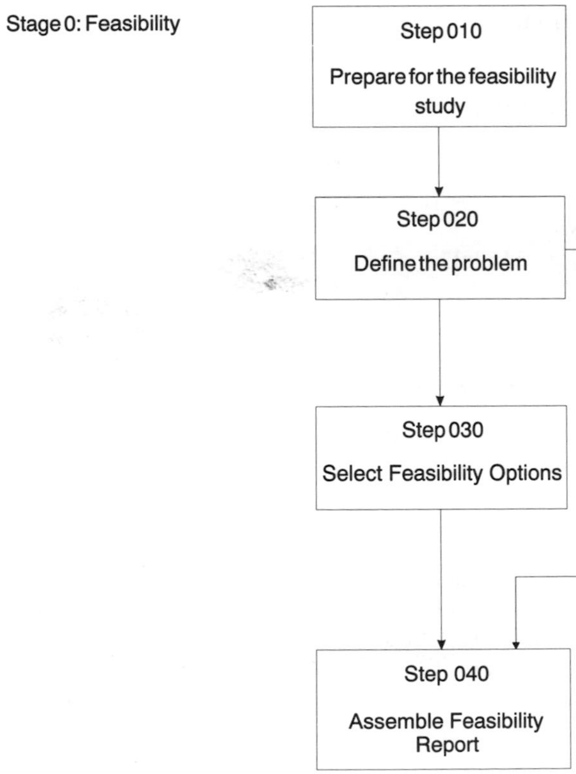

Stage 0: Feasibility

Step 010

Prepare for the feasibility study

Step 020

Define the problem

Step 030

Select Feasibility Options

Step 040

Assemble Feasibility Report

Figure 2.2 Schematic of Feasibility Study Module

2.3 Requirements Analysis Module

The purpose of the Requirements Analysis Module is to define in overview the requirements for the new system by extracting the essential functionality and data from the current system and documenting the users' requirements for the new system.

Overview

The Requirements Analysis Module is the start of a 'full study' project and is therefore initiated by the establishment of the analysis framework and formulation of plans (often performed in practice as part of the project initiation).

The requirements are analysed in three related streams:

- investigation of current functionality, data, and system users;
- elicitation of requirements for the new system;
- formulation of broad system options for meeting the requirements.

The first two activities are undertaken in Stage 1: Investigation of current environment and the last activity is undertaken in Stage 2: Business System Options. Figure 2.3 gives an overview of the steps within Stages 1 and 2.

Products

DELIVERABLE END-PRODUCT: ANALYSIS OF REQUIREMENTS

- Current Services Description.
- Requirements Catalogue.
- User Catalogue.
- Selected Business System Option.

INTERMEDIATE PRODUCTS

- Business System Options.
- Current Physical Data Flow Model.

2.4 Requirements Specification Module

The purpose of the Requirements Specification Module is to analyse in detail the requirements for the new system within the framework defined by Business System Options.

Overview

The Requirements Specification Module 'fleshes out' the overview of the system defined in Business System Options by using a number of cross-checking techniques to model various different aspects of the system. Within this module, the data and functions of the system are specified in detail and the system objectives quantified. The use of specification prototyping is optionally included to assist in the definition of functions and determine broad objectives of the human–computer interface design.

Figure 2.4 gives an overview of the steps within Stage 3: Definition of Requirements.

Stage 1: Investigation of Current Environment

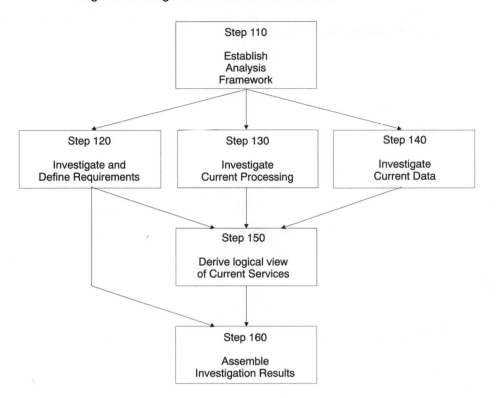

Stage 2: Business System Options

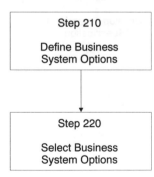

Figure 2.3   Schematic of Requirements Analysis Module

Products

DELIVERABLE END-PRODUCT: REQUIREMENTS SPECIFICATION

- Data Catalogue.
- Requirements Catalogue.
- Processing Specification.

Stage 3: Definition of Requirements

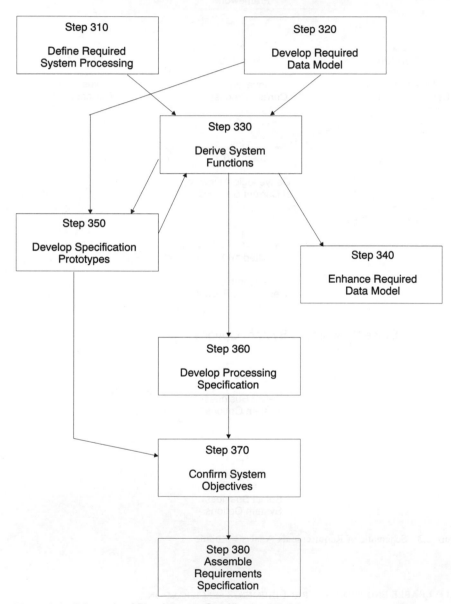

Figure 2.4 Schematic of Requirements Specification Module

INTERMEDIATE PRODUCTS

- Required System Data Flow Model.
- Event/Entity Matrix.
- Relational Data Analysis Working Paper.
- User Roles.
- Prototyping Products.

2.5 Logical System Specification Module

The purpose of the Logical System Specification Module is to provide a detailed specification of the processing and dialogue requirements for the new system and to describe the technical environment of the new system.

Overview

The Logical System Specification Module consists of two parallel streams:

- the evaluation, selection and documentation of the technical environment;
- the detailed structured specification of dialogues and processes.

Figure 2.5 gives an overview of the steps within Stage 4: Technical System Options and Stage 5: Logical Design.

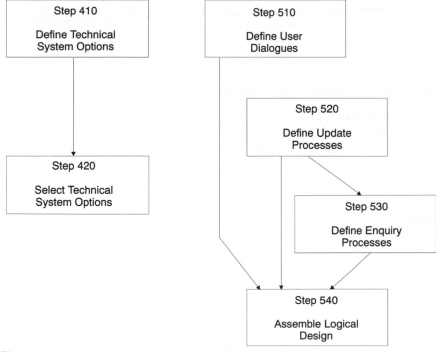

Figure 2.5 Schematic of Logical System Specification Module

Products

DELIVERABLE END-PRODUCT: LOGICAL SYSTEM SPECIFICATION

- Selected Technical System Option.
- Technical Environment Description.
- Logical Design.

INTERMEDIATE PRODUCTS

- Technical System Options.

2.6 Physical Design Module

The purpose of the Physical Design Module is to act as a bridge between the logical design and the construction phase of a project.

Overview

The inputs to physical design come from a variety of sources:

- The functionality, data, and volumetrics are derived from the SSADM documentation.
- The methods of physical design and optimization depend upon the implementation hardware and software.
- The standards for system design and documentation are often imposed by the organization.

Physical Design brings these strands together and provides techniques to map logical design products onto the physical implementation products required.

Figure 2.6 gives an overview of the steps within Stage 6: Physical Design.

Products

DELIVERABLE END-PRODUCT: PHYSICAL SYSTEM SPECIFICATION

- Physical Design.
- Application Development Standards.
- Physical Environment Specification.

2.7 Comparison with Version 3

Version 3 of SSADM has only two levels within the structural standards, namely *stages* and *steps*. Version 3 includes direct dependencies between the stages. Version 4 adds the higher level of modules into the structure of the method and clearly associates sets of end-products as deliverables from the modules. Both versions have six stages but they are partitioned differently, as shown in Table 2.2. The specific changes made in the transition from Version 3 to Version 4 are as follows:

- Version 3 Stage 2 is split into two stages.
- Relational Data Analysis is brought forward into Requirements Specification (Step 340) instead of being part of Logical Design.

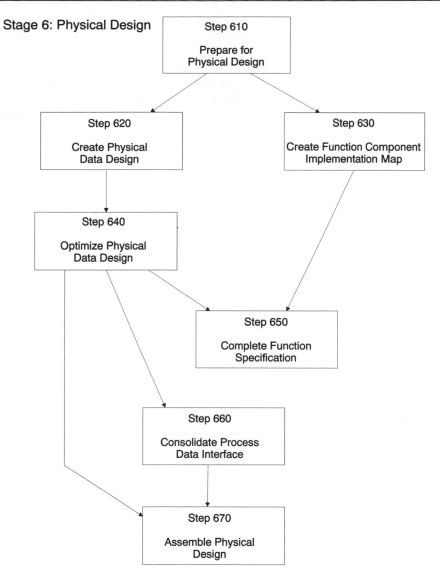

Figure 2.6 Schematic of Physical Design Module

Table 2.2

Version 3 stage	Version 4 stage
1 Analysis of system operations common problems	1 Investigation of current environment
2 Specification of Requirements	2 Business System Options
	3 Definition of Requirements
3 Selection of Technical Options	4 Technical System Options
4 Data design	
5 Process design	5 Logical Design
6 Physical Design	6 Physical Design

- There is no separate Logical Data Design stage in Version 4.
- Current System Logical Data Flow Diagrams are moved from Version 3 Stage 2 to Version 4 Stage 1.
- Specification Prototyping is formally introduced as a step in Version 4.
- Technical System Options is in parallel with Logical Design in Version 4 where Technical Options preceded Logical Design in Version 3.
- Dialogue Design is part of Logical Design in Version 4 where Dialogues were designed during the Specification of Requirements in Version 3.

3. Logical Data Modelling

3.1 Introduction

Data is modelled in the earlier stages of SSADM in order to:

- understand clearly the information (data) that underlies the current system and its interrelationships;
- build the requirements for data onto this basic model;
- incorporate all the data items and access paths required for the new system's processing;
- act as a basis for the database or file design of the implemented system.

Two main techniques are used to model data:

- Logical Data Modelling (LDM), which builds a model 'top-down' from the current information and requirements.
- Relational Data Analysis (RDA), which builds a model 'bottom-up' from the data items used as input or output from the processing of the system.

3.2 Naming conventions

Version 4 naming

The term *Logical Data Model* (LDM) defines the complete set of documentation describing the data structure of the system. The set consists of:

- the Logical Data Structure diagram;
- Entity Descriptions;
- Relationship Descriptions.

The Logical Data Structure diagram contains only two elements:

- *entities*, which represent the logical groupings of data identified by a unique key;
- *relationships*, each of which represents the precise way in which an entity relates to another entity.

There is only one type of entity but there are several types of relationship. Relationships can be *one-to-one* (1:1), *one-to-many* (1:m), or *many-to-many* (m:n). Relationships can be *optional* or *mandatory* and can be exclusive to other relationships shown using an *exclusive arc*. Each relationship end has a *relationship identifier*, which is used as the basis for *relationship statements*. *Relationship Descriptions* include the relationship statements and volumetric information about the relationships.

19

Three different Logical Data Models are produced within the stages of SSADM: *Overview LDM*, the *Current Environment LDM* and the *Required System LDM*.

The Overview LDM is developed early in the analysis. It helps gain an initial understanding of the system and is not described in detail. In most cases, therefore, it will only consist of an *Overview Logical Data Structure*.

The *Data Catalogue* is a central description of all *attributes* or *data items* within the system. This consists of:

- Attribute/Data Item Descriptions;
- Grouped Domain Descriptions.

3.3 Place in structure

The techniques of data modelling are used primarily in the Requirements Analysis and Requirements Specification phases of Version 4. However, the Required System Logical Data Model may also be updated in the Logical System Specification Phase.

Table 3.1 shows the steps in SSADM that involve the use of data modelling techniques (LDM or RDA) or update the Logical Data Model (possibly as part of another activity).

3.4 Place in Product Breakdown Structure

The Overview Logical Data Model does not appear in the Product Breakdown Structure as it is transferred into the Current Environment Logical Data Model before the end of the Requirements Analysis Module. As indicated above, the Required System Logical Data Model is created in the Requirements Specification Module but may be updated in the Logical System Specification Module. Thus, Logical Data Models and the Data Catalogue appear in three places in the Product Breakdown Structure as shown in Fig. 3.1.

The Data Catalogue consists of:

- Attribute/Data Item Descriptions;
- Grouped Domain Descriptions.

The Logical Data Models consist of:

- Logical Data Structure;
- Entity Descriptions;
- Relationship Descriptions.

Table 3.1 Logical Data Modelling in the SSADM structure

Step	LDM technique	RDA technique	Type[a]	Created/Amended
110	Y		O	C
140	Y	(Y)	C	C
320	Y	(Y)	R	C
340		Y	R	A
360			R	A
520			R	A

[a]O, Overview LDM; C, Current Environment LDM; R, Required System LDM.

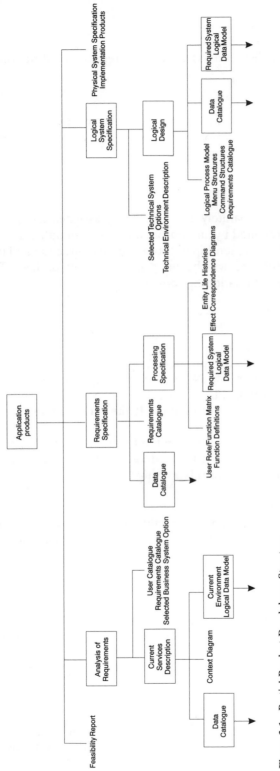

Figure 3.1 Partial Product Breakdown Structure

21

3.5 Notation and use

Entity representation

Soft boxes (rectangles with rounded corners) are used to represent entities on a Logical Data Structure Diagram.

Relationship representation

Relationships between entities are drawn as lines with 'crow's feet' to represent the degree of the relationship. Each relationship is between two entities and can be considered as having two halves:

- from the first entity to the second entity;
- from the second entity to the first entity.

Each half should be considered separately and optionality assigned independently. A mandatory relationship is represented by an unbroken line and an optional relationship is shown using a dashed line. If a relationship is optional in one direction and mandatory in the other direction, half the line will be unbroken and half dashed. There are, therefore, three different types of relationship possible.

- half optional, half mandatory;
- totally mandatory;
- totally optional.

Relationships can be 1:1, 1:m, or m:n for the Overview and Current Environment Logical Data Models. However, after Step 340 the Required System Logical Data Model should only contain 1:m relationships. Please note that it is possible to have more than one relationship between the same pair of entities. If the relationships are all *one-to-many* they must be in the same direction.

The use of optionality on relationships is illustrated by the following examples. These are based on the Required System Logical Data Model of the Opera Booking System (see Appendix F) and will only show 1:m relationships.

HALF OPTIONAL, HALF MANDATORY

For a 1:m relationship, it is possible to show optionality from the master to detail or from the detail to master. Both types are illustrated in Fig. 3.2, which shows that the relationship from 'part of theatre' to 'bookable seat' is optional, and the relationship from 'bookable seat' to 'part of theatre' is mandatory. Thus, any part of the theatre can be without any bookable seats in it (e.g. the royal box), but any bookable seat must belong to a part of the theatre.

Similarly, the relationship from 'production' to 'opera performance' is mandatory, and the relationship from 'opera performance' to 'production' is optional. This means that each production must have at least one opera performance associated with it, but it is possible to have an opera performance without a production, e.g. where there are *ad hoc* recitals by operatic performers.

TOTALLY MANDATORY

Figure 3.3 shows a relationship that is mandatory from both ends. This type of relationship is drawn with both halves of the line unbroken. Figure 3.3 shows that any

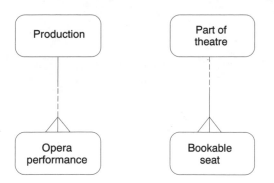

Figure 3.2 Partial Logical Data Structure showing optional/mandatory relationships

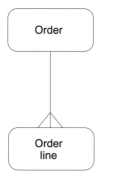

Figure 3.3 Totally mandatory relationship

instance of 'order' must be related to at least one instance of 'order line' and any instance of 'order line' must be related to one instance of 'order'. An order is therefore invalid if it does not contain at least one order line.

TOTALLY OPTIONAL

Figure 3.4 shows a relationship that is totally optional. This type is drawn with both halves of the line dashed. Figure 3.4 shows that any 'order line' may or may not be associated with one or more 'seat at performances', while a 'seat at performance' may be connected to a particular 'order line'. In the first instance not all of the order lines will immediately be satisfied and the seats are set up before being allocated to 'order lines'.

With the three different types of relationships it is possible to build the whole Logical Data Structure. The complete model for the Opera Booking System is shown in Fig. 3.5.

Relationship labelling and relationship statements

Labels are added to each end of all the relationships on the Logical Data Structure. To do this requires a very detailed knowledge of the nature of each of the relationships.

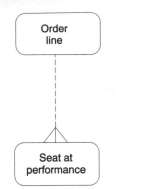

Figure 3.4 Totally optional relationship

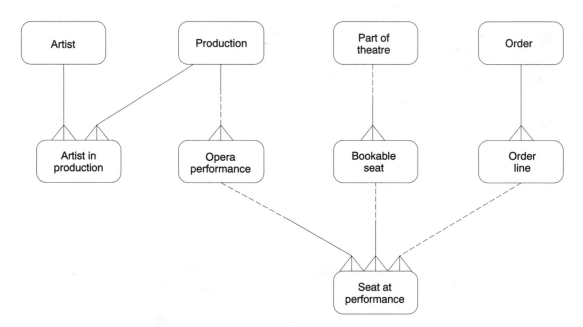

Figure 3.5 Logical Data Structure for the Opera Booking System

Each relationship has two labels:

- one describing the relationship from the first entity to the second entity;
- the other describing the relationship from the second entity to the first entity.

The purpose of the labels is to allow the relationship to be read in either direction to give a clear understanding of the precise relationship between the two entities involved. The reading of relationships is done by fitting the entity names and relationship labels into a standard 'template', the general form of which is:

Each ⟨Name of entity at one end⟩ ⟨link phrase⟩ ⟨name of entity at other end⟩

where the 'link phrase' contains one of the relationship labels. The resulting phrase is called a *relationship statement*.

For each relationship, there are two relationship statements. One statement is read from the first entity to the second entity. The other is read from the second entity to the first.

The relationship statement is composed the following way:

- The word 'Each' is the first word to reflect the fact that the statement is for a single occurrence of the first entity (even if it is at the 'many' end of a relationship).
- The name of the first entity is added.
- The words 'may be' or 'must be' are added next depending on the following criteria:
  — If the relationship is mandatory from the first entity to the second (i.e. if the line is continuous at the end next to the first entity), the expression *must be* is used.
  — If the relationship is optional from the first entity to the second (i.e. if the line is dashed at the end next to the first entity) the expression *may be* is used.
- The relationship label at the end nearest the first entity is added.
- The words 'one or more' or 'one and only one' are added next depending on the following criteria:
  — If there is a 'crow's foot' at the end nearest the second entity, the words 'one or more' are used.
  — If there is no 'crow's foot' at the end nearest the second entity, the words 'one and only one' are used.
- The name of the second entity is added (in the plural if the words 'one or more' precede it).

This is best illustrated in terms of an example. In Fig. 3.2, the relationship between 'production' and 'opera performance' is mandatory from the 'production' end and optional from the 'opera performance' end. The labels 'performed at' and 'for' can be added as shown in Fig. 3.6. Thus a production is 'performed at' an opera performance and an opera performance is 'for' a production.

The two relationship statements for this relationship are built up as follows:

1. From the 'production' end of the relationship, the relationship is mandatory, so the words 'must be' will be used. There is a 'crow's foot' at the 'opera performance' end

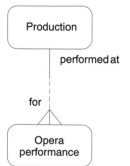

Figure 3.6 Relationship labels

of the relationship, so the words 'one or more' are used. Therefore the relationship from the 'production' end reads:

Each 'production' must be performed at one or more 'opera performances'

2. From the 'opera performance' end of the relationship, the relationship is optional, so the words 'may be' are used. There is no crow's foot at the 'production' end of the relationship, so the words 'one and only one' are used. Therefore the relationship from the 'opera performance' end reads:

Each 'opera performance' may be for one and only one 'production'

The Logical Data Structure for the Opera Booking System with all the relationship labels added is shown in Fig. 3.7. The complete set of relationship statements for this Logical Data Structure are as follows:

Each 'order line' must be present on one and only one 'order'.
Each 'order' must be composed of one or more 'order lines'.
Each 'order line' may be satisfied by one or more 'seat at performances'
Each 'seat at performance' may be assigned to one and only one 'order line'.
Each 'part of theatre' may be containing one or more 'bookable seats'.
Each 'bookable seat' must be sited in one and only one 'part of theatre'.
Each 'bookable seat' may be sold as one or more 'seat at performances'.
Each 'seat at performance' must be for one and only one 'bookable seat'.

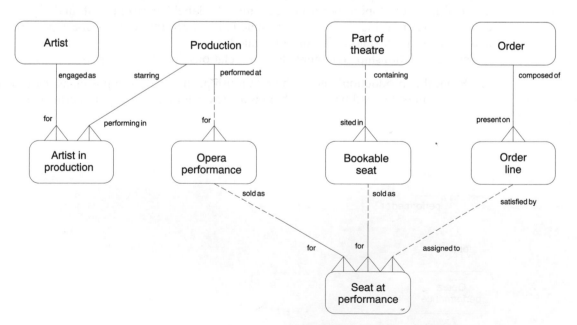

Figure 3.7 Logical Data Structure for the Opera Booking System showing relationship labels

Each 'production' must be performed at one or more 'opera performances'.
Each 'opera performance' may be for one and only one 'production'.
Each 'opera performance' may be sold as one or more 'seat at performances'.
Each 'seat at performance' must be for one and only one 'opera performance'.
Each 'production' must be starring one or more 'artist in productions'.
Each 'artist in production' must be performing in one and only one 'production'.
Each 'artist' must be engaged as one or more 'artist in production'.
Each 'artist in production' must be for one and only one 'artist'.

When choosing relationship labels it is important to keep in mind the relationship statements, which must be as grammatically correct as possible.

The relationship statements are a very useful aid to checking that the Logical Data Structure has been drawn correctly. If users have difficulty in understanding the diagram, they should be able to validate the Logical Data Structure by simply validating the relationship statements.

Exclusive relationships

Where two or more relationships are mutually exclusive this is shown on the Logical Data Structure by drawing a continuous arc across all the mutually exclusive relationships. This indicates that if one relationship in the set bounded by the arc exists, then the others in the same set cannot exist.

Each exclusive arc can only span relationships that are either all mandatory or all optional. While it is possible that a relationship can appear in more than one exclusive set, it is preferable, for clarity, that each relationship should be restricted to appearing in only one exclusive set. Otherwise, it becomes very complicated to work out the 'rules' for which relationships can coexist.

It is possible to identify 'sub-types' of entities by the use of an exclusive arc. A sub-type is where an entity can contain different attributes or have different relationships depending upon the circumstances. An example might be that 'seat at performance' is either a 'seat at performance by order' or a 'seat at performance in person'. This can be represented as shown in Fig. 3.8.

If the Logical Data Structure were to be redrawn to show this, the relationship to 'order line' would be from the 'by order' sub-type and be mandatory, rather than from the entity super-type 'seat at performance'.

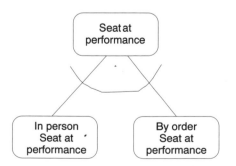

Figure 3.8 Exclusive arc showing sub-types

Recursive relationships

It is possible to indicate that an entity is a hierarchy by the use of a recursive relationship, often called a 'pig's ear'. This is represented by a fully optional *one-to-many* relationship starting and ending at the same entity. For example, it would be possible for 'part of theatre' to represent large areas of the theatre or sub-parts of those same areas.

Partitioning of models

In constructing a Logical Data Structure, it is sometimes useful to break the structure down into smaller parts, each substructure covering a discrete area of the total system.

Such partitioning will result in some relationships from entities at the edge of the substructures being excluded as the related entities are in another substructure. If so, the entities at the edge are described as 'incomplete'. Any incomplete entity is shown on the structure as a dashed box.

In a similar way, it is possible that an exclusive arc is 'incomplete' because the other relationships in the exclusive set are not in this particular substructure. Figure 3.9 shows an example of an incomplete entity and incomplete exclusive relationship. In this example both the entity 'bookable seat' and the exclusive arc are incomplete and are therefore drawn using dashed lines. Any entity drawn under an incomplete exclusive arc must always be incomplete itself.

Relational Data Analysis

Relational Data Analysis is a widely used technique for building normalized relations from sets of unstructured data based on the work of Codd and documented fully by Date. In SSADM, Relational Data Analysis is used to supplement the Logical Data Modelling technique to check that:

- all attributes (except derived or transient attributes) required by the processing are held within the Logical Data Model;
- the Logical Data Model is a normalized view of the data;
- the Logical Data Model is flexible.

The two data modelling techniques are complementary in that the Logical Data Model is first developed 'top-down' from the statement of the users' requirements. The

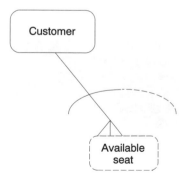

Figure 3.9 Example of incomplete entity and exclusive relationship

Relational Data Analysis technique builds a data model 'bottom-up', working from sets of attributes which are then logically grouped. By building the results of Relational Data Analysis into the Logical Data Model, the advantages of both techniques can be realized and cross-checked against one another.

Relational Data Analysis is done in a number of steps leading from a set of un-normalized data through first normal form, second normal form, and third normal form. (In exceptional circumstances, fourth and fifth normal form may be considered useful. However, these are not covered here as the majority of projects will only require the first three to be carried out.)

INPUTS TO RDA

The un-normalized data would principally be derived from the I/O Structures developed as part of the Function Definition. Each I/O Structure may be considered as a candidate for Relational Data Analysis (not all I/O Structures need to be used if duplicate or overlapping data is present — a subset should be selected which covers the entire data content of the system).

Forms and reports may also be used as an input to RDA. If the technique is used as part of the investigation of the current environment, documents from the current system will be used. If the technique is used in the Definition of Requirements, any documentation that will be part of the new system will be used.

UN-NORMALIZED DATA

The un-normalized data may be listed from the source of the data. If the data is already structured (as in the I/O Structure), the un-normalized data may reflect this structuring. An overall key is chosen for the un-normalized relation. This can be a single data item or a group of data items. In general, the key chosen should:

- only have a single value for the set of un-normalized data;
- be as simple as possible;
- preferably be a code or number.

FIRST NORMAL FORM

The un-normalized data is converted to first normal form by identifying repeating groups of data. A repeating group is a set of data items that can have more than one value for a given value of the prime key of the un-normalized set. For each repeating group, a supplementary key is identified which, when added to the original key chosen for the un-normalized set, uniquely identifies each occurrence of the group. Each group is then a first normal form relation. Many data items are not part of repeating groups and these are left under their original key.

SECOND NORMAL FORM

Once the analyst has made sure that all data dependencies are understood, the second normal form can be derived from the first normal form by identifying part-key dependencies. There are two activities here, the first removes redundant attributes from the primary key, and the second removes attributes not fully dependent on the primary key. Again, any data items not altered within second normal form are simply left under their first normal form key.

THIRD NORMAL FORM

Third normal form identifies inter-data dependencies in the second normal form relations. This time, instead of only looking for dependencies in the key, all the data items within a relation are examined to find out if interdependencies exist between them. Thus, given one of the data items within the relation, there is only one possible value for one or more of the other data items within the same relation. Any such interdependent data items are then grouped into new relations. The key of the new relation remains in the original relation and is marked as a 'foreign key', which acts as a pointer to the new relation.

OPTIMIZATION

All the resulting third normal form relations are compared with each other and any duplicates removed. After careful consideration, any relations with the same key are merged.

Example of third normal form

Figure 3.10 shows an order form for subscribers from the Opera Booking System. This is from the current system, but a decision has been made that the design of the booking form will remain the same for the new system and so it is used as input to Relational Data Analysis during the Definition of Requirements. (It is always preferable to use a filled-in

Booking form for opera series — Summer 1992

Name: R. Wagner   Address: 22 Valkyrie Close   Subscriber no. 1234
                             Valhalla
                             Oxfordshire
                             OX9 3BD

Date of Performance	Title of Opera	Part of theatre	No. of Seats	Price per seat	Total
13/8/92	Das Rhiengold	Stalls	2	59.45	118.90
14/8/92	Die Walkure	Stalls	2	59.45	118.90
16/8/92	Siegfried	Stalls	2	59.45	118.90
18/8/92	Gotterdammerung	Stalls	2	59.45	118.90
				Total for Order	475.60

Method of payment   Access [ ]   Visa [✓]   Credit card no: 4929 930 659 223
(Tick one box only)   Expiry date 09/96
              American Express [ ]   Cheque [ ]   Signature: R. Wagner

Figure 3.10   Booking form from Opera Booking System

form as an input to Relational Data Analysis as this will highlight any data relationships that are not always apparent from the blank form.)

The following data items are identified from the form:

Booking season
Customer name
Customer address
Subscriber number
Date of performance
Title of opera
Part of theatre
Number of seats
Price per seat
Line total
Order total
Method of payment
Credit card number
Credit card expiry date

The derived data items 'line total' and 'order total' do not strictly need to be included here as they are not elemental items and depend only on other fields on the order. However, in this case, they are included as these are totals entered by the customer and may be required in case of any discrepancies or disputes.

The I/O Structure for the function 'receive subscriber's order' is also used in the RDA exercise. This is shown in Fig. 3.11. Booking season is included because more that one season's data is held on the system.

Figure 3.12 shows the RDA Working Paper completed for the order. Here, the key identified for the un-normalized form (UNF column) is a combination of 'subscriber number' and 'booking season'. Taken in the context of a single form, the 'subscriber number' would be a unique key. However, taken in conjunction with other forms, this requires the 'booking season' for uniqueness.

The UNF level is derived from the I/O Structure in Fig. 3.11. Level 1 reflects non-repeating items and level 2 reflects all items repeating together. This helps in identifying the repeating groups for first normal form.

In first normal form (1NF) only one repeating group is identified, corresponding to the order lines. As there is no unique identifier for a line on the form, a derived item 'line number' is introduced. 'Line number' is not unique in its own right so it is coupled together with the UNF key to form a *composite* (or *hierarchic*) *key* that we have indicated by the use of brackets although this is not strictly required by the method. The brackets denote that the 'line number' will never be used separately from the qualifiers 'booking season' and 'subscriber number'.

At second normal form (2NF), a part-key dependency is identified, 'customer name' and 'customer address' depend only on the 'subscriber number'. This is pulled out as a separate relation.

No inter-data dependencies are identified at third normal form (3NF), 'credit card number' and 'credit card expiry date' are related but do not have any direct dependency.

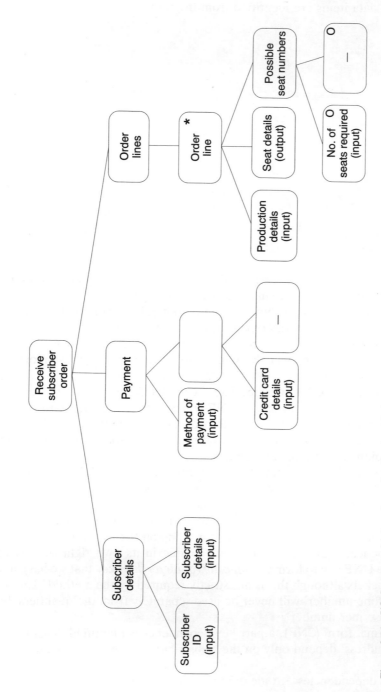

Figure 3.11 I/O Structure for 'receive subscriber's order'

UNF attributes	UNF level	1NF	2NF	3NF	Result relation	Result attributes
Customer name Customer address Subscriber number Date of performance Title of opera Part of theatre No. of seats Price per seat Line total Order total Method of payment Credit card number Credit card expiry date Booking season	1 1 1 2 2 2 2 2 2 1 1 1 1 1	Booking season Subscriber number Customer name Customer address Method of payment Credit card number Credit card expiry date Order total { Booking season Subscriber number Line number Date of performance Title of opera Part of theatre No. of seats Price per seat Line total	Booking season Subscriber number Method of payment Credit card number Credit card expiry date Order total Subscriber number Customer name Customer address	Booking season Subscriber number Method of payment Credit card number Credit card expiry date Order total Subscriber number Customer name Customer address { Booking season Subscriber number Line number Date of performance Title of opera Part of theatre No. of seats Price per seat Line total	Order Subscriber Order line	Booking season Subscriber number Method of payment Credit card number Credit card expiry date Order total Subscriber number Customer name Customer address { Booking season Subscriber number Line number Date of performance Title of opera Part of theatre No. of seats Price per seat Line total

Figure 3.12 Completed RDA Working Paper for the order form

The resulting third normal form relations are:

Order   *Booking season*
          Subscriber number
          Method of payment
          Credit card number
          Credit card expiry date
          Order total

Subscriber *Subscriber number*
          Customer name
          Customer address

Order line *(Booking season)*
          (Subscriber number)
          (Line number)
          Date of performance
          Title of opera
          Part of theatre
          No. of seats
          Price per seat
          Line total

These three relations are candidate entities for the Required System Logical Data Model. Figure 3.13 shows the Entity Descriptions from the Logical Data Model after comparison with the results of Relational Data Analysis. The Relational Data Analysis exercise has identified several areas that were not included in the Logical Data Structure:

- There is no 'subscriber' entity.
- The 'order' entity does not contain details of the method of payment.
- The key of 'Order' was replaced by a simple 'order number'.
- In a number of cases there were differences between the naming standards adopted. Generally the ones from the pre-RDA structure were used.
- The totals were not included.

Because this is only a partial exercise, a number of items have been identified on the order line that will not be held, such as 'title of opera', and this must be resolved once the total Relational Data Analysis exercise is complete.

3.6 Comparison with Version 3

The use and meaning of Logical Data Modelling has not changed much from the Version 3 Logical Data Structuring. However, there are several differences worth highlighting. These may be summarized under the following headings:

- Notation for entities and relationships.
- Labelling of relationships and relationship statements.
- Naming of data structures.

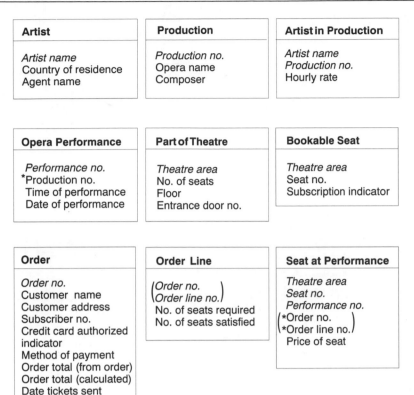

Artist	Production	Artist in Production
Artist name Country of residence Agent name	*Production no.* Opera name Composer	*Artist name* *Production no.* Hourly rate

Opera Performance	Part of Theatre	Bookable Seat
Performance no. *Production no.* Time of performance Date of performance	*Theatre area* No. of seats Floor Entrance door no.	*Theatre area* Seat no. Subscription indicator

Order	Order Line	Seat at Performance
Order no. Customer name Customer address Subscriber no. Credit card authorized indicator Method of payment Order total (from order) Order total (calculated) Date tickets sent	(*Order no.* *Order line no.*) No. of seats required No. of seats satisfied	*Theatre area* *Seat no.* *Performance no.* (*Order no. *Order line no.) Price of seat

Figure 3.13 Entity Descriptions

- Profile of Relational Data Analysis.
- Composite logical data design.
- Operational masters.

Notation for entities and relationships

The notational changes can best be described by direct comparison between a Logical Data Structure drawn to the Version 3 standard and the same diagram drawn to Version 4 standards.

Figure 3.14 shows the Logical Data Structure developed for the Opera Booking System as part of the analysis of the current system. It is drawn to Version 3 standards. This should be compared with Fig. 3.7 above, which is the equivalent Version 4 structure. The differences are as follows.

ENTITY REPRESENTATION

In SSADM Version 3 'hard boxes' are used to represent entities; in Version 4 'soft boxes' are used.

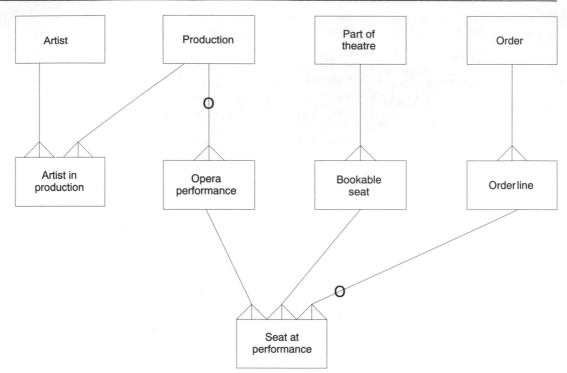

Figure 3.14 Version 3 Logical Data Structure

RELATIONSHIP REPRESENTATION

In Fig. 3.12 there are two types of relationship represented: the standard one-to-many (1:m) relationship and the optional one-to-many relationship. These are the only relationship types supported in Version 3. They can be defined as follows:

- the mandatory 1:m relationship, shown using the unbroken line with the crow's foot. This is drawn between the two entities such that the crow's foot is at the 'many' end. This represents a 1 to 0, 1, or many from master to detail, and a 1 to 1 from detail to master.

- the optional 1:m relationship, drawn as for the mandatory relationship with the addition of a circle to show that the relationship is optional from detail to master (this is illustrated by the relationship between 'seat at performance' and 'order line'). This represents a 1 to 0, 1, or many from master to detail, and a 1 to 0 or 1 from detail to master.

Labelling of relationships and relationship statements

The addition of labels, or names, to relationships in Version 3 is used when the nature of a relationship requires further explanation or to help distinguish relationships when there is more than one relationship between the same pair of entities. In general, only one name or label is added to the relationship and there are no rules concerning the format of the name or label.

By comparison, Version 4 adds strictly defined labels to each end of all the relationships on the Logical Data Structure. To enable the analyst to undertake this will require a very detailed knowledge of the nature of each of the relationships thus concentrating further the level of understanding of the system being documented. In addition naming the relationships increases the amount of information that is present on the Logical Data Structure. Although this change makes the technique look more 'difficult', experience has shown that the reduction in ambiguity ultimately makes data structures drawn to these standards easier to construct and understand.

Naming of data structures

The differences between the names are summarized here:

Version 3	Version 4
Current System LDS	Current Environment LDM
Required LDS	Required System LDM
Composite Logical Data Design	'Enhanced' Required System LDM

Version 3/Version 4 naming comparison

The differences between Version 3 and Version 4 naming are as follows:

Version 3	Version 4
Logical Data Structuring	Logical Data Modelling
No equivalent	Logical Data Model (LDM)
Logical Data Structure	Logical Data Structure (LDS)
Entity Descriptions	Entity Descriptions
Part of the Data Catalogue	Attribute Descriptions
No equivalent	Grouped Domain Descriptions
Less formal Relationship Descriptions	Relationship Descriptions
Entity	Entity
Relationship	Relationship
Only 1:m allowed	1:1, 1:m, m:n
Informal relationship labels	Relationship identifier
No equivalent	Relationshp statements
Operational Masters	No equivalent
Optional, mandatory, exclusive	Optional, mandatory, exclusive
Overview LDS	Overview LDM
Current LDS + support documentation	Current Environment LDM
Required LDS + support documentation	Required System LDM

3.7 Relational Data Analysis

Relational Data Analysis is a major part of the Logical Data Design stage (Stage 4) in Version 3. In practice, however, this has been seen to give too high a profile to the technique. In many cases, projects have spent a long time on Relational Data Analysis for very little gain. Therefore, Version 4 reduces the profile of Relational Data Analysis as a technique to check the completeness of the Required System Logical Data Model. In addition, in Version 3 it was necessary to produce two sets of Entity Life Histories, pre- and post-Relational Data Analysis. Within Version 4, Relational Data Analysis has been moved to an earlier point in the method, thereby reducing some of the iteration that is present in Version 3. The Relational Data Analysis technique itself has changed slightly, particularly in the area of Second Normal Form activities.

The RDA Working Paper for Version 4 is enhanced to include the UNF level based on the I/O Structure and the resulting relation names and attributes.

Composite Logical Data Design

The Composite Logical Data Design is a separate product in Version 3 produced by merging the Required Logical Data Structure and the 3NF Structure produced in Relational Data Analysis. In line with the lower profile of Relational Data Analysis, Version 4 does not distinguish the Required System Logical Data Model from the model produced as a result of merging the two structures. The merged structure is now an 'enhanced' Required System Logical Data Model.

Operational Masters

Operational Masters appear on Version 3 Logical Data Structures to indicate access requirements to entities using data items other than the prime key. This information is contained in the Effect Correspondence Diagrams and Enquiry Access Paths in Version 4 and Operational Masters do not appear on the Logical Data Model.

4. Data Flow Modelling

4.1 Introduction

Data Flow Modelling is used in the early stages of SSADM in order to:

- understand clearly the flows of data around a system;
- define the processes that transform or manipulate the data;
- identify the immediate sources and recipients of data outside the system;
- show where data is held within the system;
- act as a means of communication between analyst and user;
- form the basis of function definition and event identification.

The system represented by a Data Flow Model can be a computer system, a manual system, or an abstract view of what is currently done or required to be done.

4.2 Naming conventions

The term *Data Flow Model* (DFM) defines the complete set of documentation supporting a set of Data Flow Diagrams (DFD). Three versions of the Data Flow Model are produced:

- Current Physical Data Flow Model.
- Logical Data Flow Model.
- Required System Data Flow Model.

In addition, a set of Data Flow Diagrams may be produced in support of Business System Options.

The Data Flow Model consists of the following:

- Data Flow Diagrams (level 1 and below).
- Elementary Process Descriptions.
- External Entity Descriptions.
- I/O Descriptions.

Data Flow Diagrams consist of the following elements:

- *Processes*, which represent the transformation or manipulation of data.
- *Data flows*, which are 'channels' between other elements down which predefined sets of data may flow.
- *Data stores*, which represent data at rest. They can be subdivided into the following two types:

— *Main*, which represent groups of entities from the Logical Data Model.
— *Transient*, which represent data that is required to be held for processing and is deleted after it is used.

For the Current Physical Data Flow Model only, these can be further subdivided into the classifications:

— *Manual*, which are non-computerized stores of data.
— *Computer*, which are computerized stores.
● *External entities*, which are sources or recipients of data immediately outside the system boundary.

In the early stages of analysis, a *Context Diagram* may be drawn. This shows the entire system as a single process with all flows to and from external entities. This type of diagram has no data stores.

If the current system consists of flows of physical resources without corresponding data flows (for example in a warehouse) the Current Physical Data Flow Diagrams may be drawn as *Resource Flow Diagrams* with *Resource Flows* and *Resource Stores*.

As a start-up to the Current System Data Flow Diagrams it may be useful to draw a *Document Flow Diagram*, which shows the flows of documents between the different areas of the system. This type of diagram does not have processes or data stores.

Each process on a Data Flow Diagram may be re-expressed as a lower-level Data Flow Diagram. The set of Data Flow Diagrams is hierarchic and may be drawn to any number of levels, but it is sensible to limit this to two or three. Each process at the bottom level (i.e. which is not further decomposed) is described in textual form using an *Elementary Process Description*.

A *Data Store/Entity Cross-reference* relates the data stores of the Data Flow Model to the entities from the corresponding Logical Data Model.

4.3 Place in structure

Data Flow Modelling is used in two modules of SSADM Version 4:

● Requirements Analysis;
● Requirements Specification.

Table 4.1 shows the steps that involve the Data Flow Models.

Table 4.1

Step	Type[a]	DFM Created (C)/Amended (A)/Input (I)
100	C	C
130	C	A
150	L	C
210	B	C (optional)
220	B	A (optional)
310	R	C
330	R	I

[a]C, Current Physical Data Flow Model; L, Logical Data Flow Model; B, Business System Options Data Flow Diagrams; R, Required System Data Flow Model.

4.4 Place in Product Breakdown Structure

Data Flow Diagrams appear in only one area of the overall Product Breakdown Structure even though three (possibly four) Data Flow Models are produced. The partial Product Breakdown Structure in Fig. 4.1 shows the Logical Data Flow Model appearing in the Product Breakdown Structure for the Requirements Analysis Module only.

The reason for a Data Flow Model appearing only once is that the Product Breakdown Structure represents 'deliverables' from modules. Only one Data Flow Model is a formal end-of-module deliverable — the Logical DFM, otherwise:

- The Current Physical Data Flow Model is superseded by the Logical Data Flow Model before the end of the Requirements Analysis Module.
- The Business System Options Data Flow Diagrams are informal products and form part of the Selected Business System Option.
- The Required System Data Flow Model is superseded by the functions and events before the end of the Requirements Specification Module.

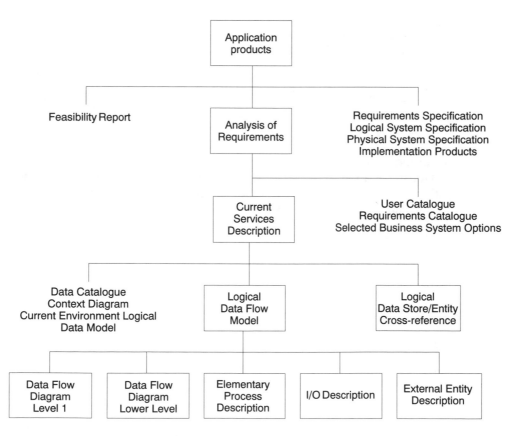

Figure 4.1 Partial Product Breakdown Structure

4.5   Notation and use

Current Physical Data Flow Model

The Current Physical Data Flow Model is a direct representation of the data flows and processes of the current system. This is developed in advance of the corresponding logical view for several reasons:

- It is possible to miss constraints or limitations imposed on the current system which will still be present in the new system (e.g. the use of standard input or output documents).
- There may be problems in the current system that should be corrected by the new system; these need to be identified before they can be resolved.
- The users often find it easier to validate a diagram that reflects the actual situation rather than an abstract representation of that situation.
- Analysts often find it easier to model a situation as it actually is rather than trying to represent immediately that situation in a logical or abstract way.

Current Physical Data Flow Diagrams are developed in a number of different ways (or a combination):

- From a Context Diagram.
- Document Flow Diagrams.
- Resource Flow Diagrams.
- Directly as Data Flow Diagrams.

CONTEXT DIAGRAM

A Context Diagram from the Opera Booking System (see Appendix F) is shown in Fig. 4.2. Here, all external sources and recipients of data are shown. This helps to define the boundary of the system from the outset.

DOCUMENT FLOW DIAGRAM

A Document Flow Diagram is produced by tracing the flow of documents or forms within a system from person to person or section to section within the area under study. This includes flows into and out of the area as well as those flows within the area. The diagram can be developed by a series of partial views, either modelling individual areas or tracing flows for individual documents.

The notation for a Document Flow Diagram is shown in Fig. 4.3.

A Data Flow Diagram is developed from a Document Flow Diagram in a series of steps:

- All documents/forms used within the system are listed.
- A diagram is developed, possibly by combining a number of partial views;
- A boundary is annotated on the diagram, showing which areas are within the boundary of the investigation and which are outside.
- The areas inside the boundary are shown as processes and data stores.

RESOURCE FLOW DIAGRAM

A Resource Flow Diagram is an appropriate start-up diagram for Data Flow Diagrams when the system is tracking flows of physical 'things' rather than being principally

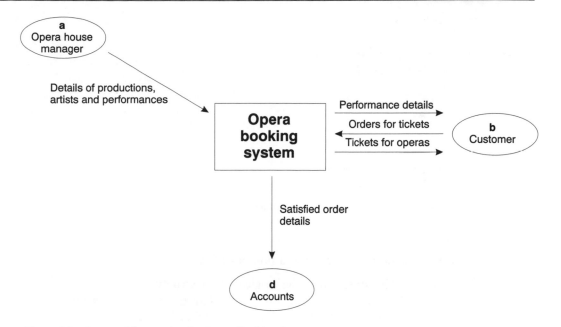

Figure 4.2 Context Diagram for the Opera Booking System

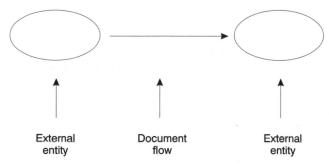

Figure 4.3 Notation for Document Flow Diagram

composed of flows of information. In some cases, there may be an implied information flow which follows the flow of physical resources. In this case, the physical resource flow can be replaced by a document/data flow. In some cases, however, there is no corresponding flow of information — this might be identified as a problem with the system! In this case, it is perfectly appropriate for physical resource flows to appear on Current Physical Data Flow Diagrams.

The notation of a Resource Flow Diagram is shown in Fig. 4.4.

Logical Data Flow Model

The Logical Data Flow Model is a logical representation of the current system. It is derived directly from the Current Physical Data Flow Diagrams using a procedure called 'logicalization'.

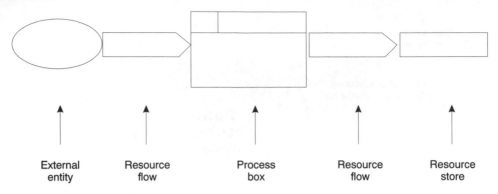

| External | Resource | Process | Resource | Resource |
| entity | flow | box | flow | store |

Figure 4.4 Notation for Resource Flow Diagram

Logicalization is achieved in a number of steps:

1. Derive data stores from the Logical Data Structure.
2. Logicalize the bottom-level processes, data flows, and transient data stores.
3. Group bottom-level processes to form higher-level processes.

Data Stores are directly cross-referenced to the entities from the Logical Data Model during logicalization. The Logical Data Store/Entity Cross-reference from the Opera Booking System is shown in Fig. 4.5.

The processes and data flows on the Logical Data Flow Diagrams are derived by removing physical aspects from the processes and data flows of the Current Physical Data Flow Diagrams. Other intermediate products that may be used during logicalization are the Process/Entity Matrix and the Logical/Physical Data Store Cross-reference.

Required System Data Flow Model

The Required System Data Flow Model is based upon the outline defined in Business System Options. In the majority of cases, the Required System Data Flow Model will be based on the Logical Data Flow Model. However, where Business System Options have identified a need for radical change, they will be developed directly from the selected BSO.

The Data Flow Diagram in Fig. 4.6 is the level 1 Required System Data Flow Diagram for the Opera Booking System. Here, the basic notation and naming conventions are as follows:

- ovals are external entities;
- the external entity identifier is a lower-case alpha character;
- rectangles are processes;
- the process identifier is a number;
- open-ended boxes are data stores;
- the data store identifier is 'D' plus a number;
- arrows are data flows;
- data flows at the bottom level are labelled with their data content.

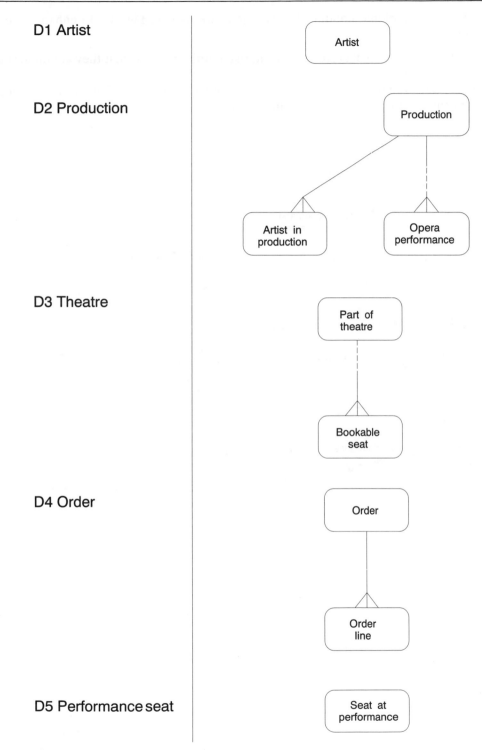

D1 Artist

D2 Production

D3 Theatre

D4 Order

D5 Performance seat

Figure 4.5 Logical Data Store/Entity Cross-reference from the Opera Booking System

Several additional notational conventions are used in Fig. 4.6 to enhance the basic notation:

- a bar is added to data stores and external entities to show that they are duplicated on this diagram;
- an asterisk in the bottom right corner of a process box shows that it is not decomposed further — there will be an Elementary Process Description for this process;

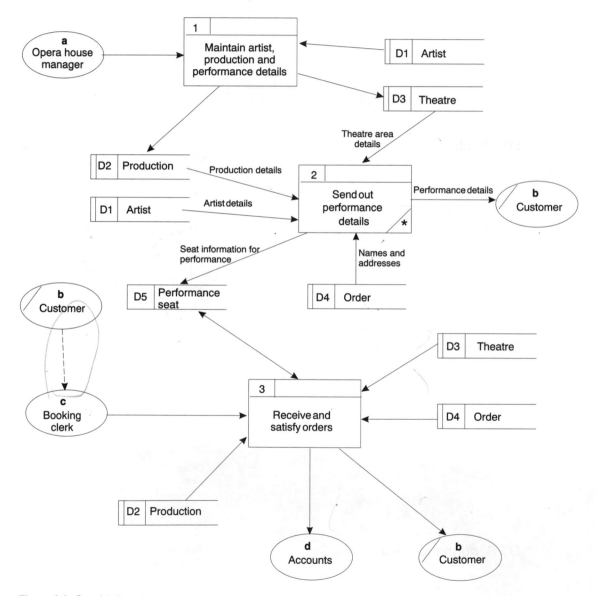

Figure 4.6 Level 1 Required System Data Flow Diagram for the Opera Booking System

- a dashed data flow between external entities is added for flows of information that are represented on the Data Flow Diagram even though they are outside the system boundary.

Figure 4.7 on page 48 shows a level 2 Data Flow Diagram which is the expansion of process 3 in Fig. 4.6. The boundary box denotes the boundary or process 3. All data flows across the boundary should correspond to (or be expansions of) flows to and from the process at the higher level. All flows are one-way only as they are at the bottom level.

Additional conventions shown in Fig. 4.7 are as follows:

- process identifiers are extensions of the owning process identifier;
- a transient data store is shown which is internal to process 3:
  — it is identified by a 'T';
  — the number of the identifier is an extension of the owning process identifier.

This level 2 DFD includes a process (3.3) that is purely a retrieval. While it is not normal to place retrievals on the DFDs, if they are important enough then they should be included. In this case the printing of tickets is regarded as a very important retrieval that is fundamental to the business and is therefore included on the DFD.

Although not illustrated in the example it is possible hierarchically to decompose both external entities and data stores. For example, a data store D1: Order could be decomposed to D1a: Subscriber order and D1b: Non-subscriber order. Similarly, an external entity with the identifier 'a' could be decomposed to 'a1' and 'a2'.

4.6 Comparison with Version 3

The changes from Version 3 to Version 4 are principally cosmetic in nature with the introduction of certain conventions that enrich the notation. The changes are in the following areas:

- naming conventions;
- place in the structure;
- external entities;
- process boxes;
- data stores;
- data flows;
- additional changes.

Naming conventions

In general the technique has changed its name from Data Flow Diagramming to Data Flow Modelling. The Data Flow Model has been introduced in Version 4. In Version 3, the equivalent was described as Data Flow Diagrams and supporting documentation.

The naming of the textual description of bottom-level processes has changed:

- Version 3 — Elementary Function Descriptions.
- Version 4 — Elementary Process Descriptions.

Physical Resource Flows are renamed Resource Flows.

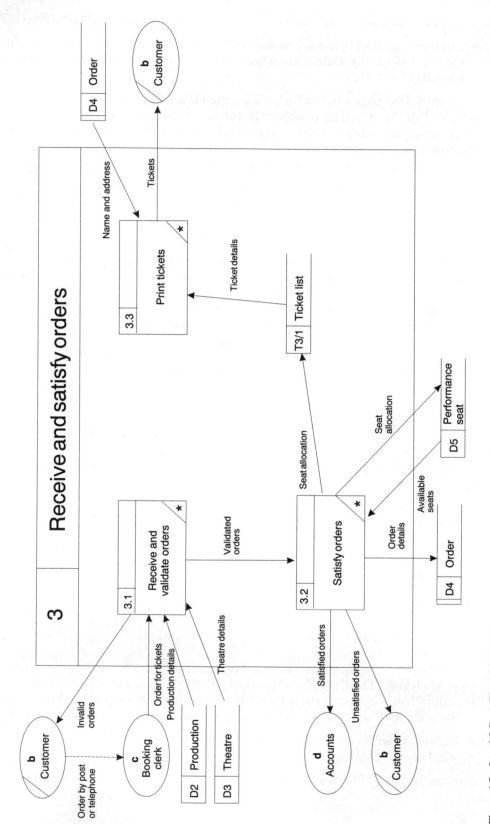

Figure 4.7 Level 2 Data Flow Diagram for process 3

Place in the structure

The change in the place of DFDs in the structure is that the logicalization of the Current Physical Data Flow Model is the final step in Stage 1 instead of being the first step in Stage 2. It is therefore the Logical DFM that becomes the product of Stage 1.

External entities

The notation for external entities has been enhanced in Version 4 to indicate duplication on a single diagram. If it is necessary to duplicate an external entity, this is indicated by a small bar across the top of the external entity symbol. This is an aid to the clarity of the diagram and does not change the use of external entities on diagrams in any way.

Process boxes

A new notation has been introduced in Version 4 to indicate bottom-level process boxes on diagrams. Bottom-level process boxes are indicated by the addition of an asterisk in the bottom right corner of the box, enclosed by a line. Again, this is an aid to clarity rather than a change in the technique. This makes clear which processes are broken down further and which processes should be described by an Elementary Process Description.

Data stores

A notation distinguishing transient data stores from main data stores is introduced in Version 4. The notation is used in all sets of Data Flow Diagrams though it is most unlikely to be found in the Current Physical Data Flow Model. The identifier of a transient data store is prefixed with a 'T' instead of a 'D'.

For the Current Physical Data Flow Model, it is possible to distinguish between manual transient data stores and automated transient data stores by the use of the 'T' prefix in conjunction with the 'M' or 'D', e.g. T1(M) or T2(D). After logicalization, manual data stores are not distinguished so all data stores are identified either as Tn or Dn where n is a number.

Data flows

In Version 4, a more formal rule is introduced to enforce good practice in the use of two-way data flows. The rule is that two-way data flows may not be connected to a bottom-level process box. In order to be completely clear about the data flowing to and from a process, at the bottom level all data flows must be one-way only.

To avoid confusing and complicated diagrams, two-way flows are allowed on higher-level Data Flow Diagrams provided that they are broken down into one-way flows in each direction at the lower levels.

Version 4 includes an extension in the use of data flows between external entities. If a flow of information between external entities is important, it may be useful to add it to the diagram. In this case, a data flow is shown as a dotted line between external entities. This is only used where the external-to-external flow is of direct relevance in 'completing' the picture. It is not intended that this should be used to start modelling manual procedures.

Additional changes

Some of the other changes in the area of Data Flow Modelling are summarized here:

- The decomposition of external entities and data stores is described in Version 4. This enables external entities and data stores to be broken down within the set of Data Flow Diagrams. This area was introduced in earlier versions but was excluded from previous manuals.
- Context Diagrams, Document Flow Diagrams and Resource Flow Diagrams are used as an initial aid in DFD construction (again, this was in earlier versions but was not in previous reference material).
- A Process/Entity matrix and Logical/Physical Cross-reference have been introduced to aid in the logicalization of the Current Physical Data Flow Model.

5. Requirements Definition

5.1 Introduction

Requirements definition is a technique that underpins much of the analysis and design activity in SSADM. It is a non-diagrammatic technique covering the collection and collation of functional and non-functional requirements. The Requirements Catalogue is used to document requirements in textual form from the outset of the project.

All functional requirements from the Requirements Catalogue are developed further using techniques that model functionality and data. Non-functional requirements are carried forward into design to define aspects of the system not covered by other SSADM techniques.

5.2 Naming conventions

The technique of *requirements definition* is used to collect and document the users' requirements from the outset of the project. The *Requirements Catalogue* documents each of the requirements for the new system and distinguishes *functional* and *non-functional* requirements.

5.3 Place in structure

Requirements definition is an iterative technique that effectively runs in parallel with most other SSADM techniques. This means that the Requirements Catalogue is involved in most of the steps in the Requirements Analysis and Requirements Specification Modules. It also acts as an important input to Logical and Physical Design.

There are only two steps that are centred on defining and updating requirements:

- Step 120: Investigate and define requirements.
- Step 370: Confirm system objectives.

However, the Requirements Catalogue may also be used as an input to, or updated as a result of using, other SSADM techniques. Therefore, the complete list of steps involving the Requirements Catalogue is extensive, as shown in Table 5.1.

5.4 Place in Product Breakdown Structure

The Requirements Catalogue appears in several places in the Product Breakdown Structure. Unlike other SSADM products, it is not given a different name in each

Table 5.1 Involvement of Requirements Catalogue in SSADM steps

Step	Requirements Catalogue entries (Created/Amended/Input)
110	C
120	I/A
130	I/A
140	I/A
150	I/A
210	I
310	I/A
320	I/A
330	I/A
350	I/A
360	I/A
370	I/A
410	I
510	I
630	I/A
640	I/A
650	I/A
660	I/A
670	I/A

context. This is because there is only one Requirements Catalogue, which is updated iteratively as the project proceeds. It can be regarded as a store into which requirements are put and then satisfied. Thus, it is possible to identify/satisfy requirements at most parts of the analysis and design stages.

The Requirements Catalogue is a constituent part of:

• Analysis of Requirements.
• Requirements Specification.
• Logical Design (which is part of the Logical System Specification).

This is illustrated in Fig. 5.1.

5.5 Notation and use

The Requirements Catalogue is created in Step 110 (unless a feasibility study has already created it, in which case it is revised in Step 110). As the investigation of the current environment proceeds, users' requirements are documented. At this stage, it is likely that the requirements will be based on problems in the current system and may not be strictly quantified. During logicalization of the Data Flow Model, physical constraints taken off the diagrams may be added to the Requirements Catalogue.

In Stage 2, the Requirements Catalogue acts as a very important input to the creation of Business System Options, though it is not updated at this time.

The development of the Required System Data Flow Model and Required System Logical Data Model may result in the Requirements Catalogue being updated in the light of chosen BSO as some requirements may be discarded at this point. Requirements satisfied by Function Definition, Specification Prototyping and Entity/Event Modelling

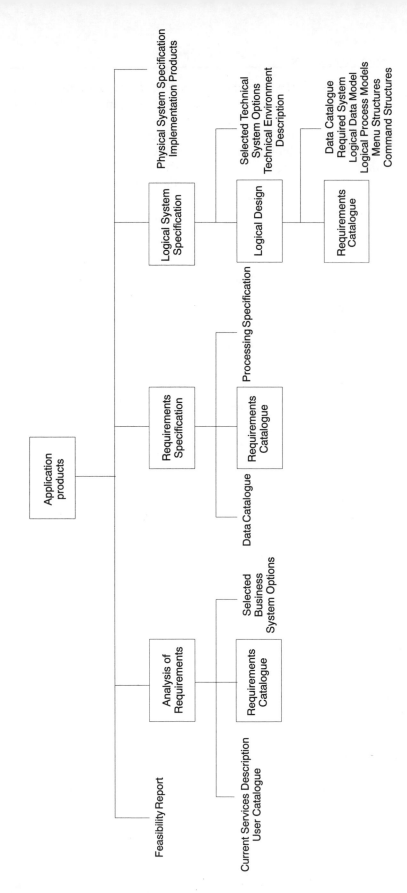

Figure 5.1 Partial Product Breakdown Structure

are cross-referred as they are developed. In Step 370, the Requirements Catalogue is checked to ensure that all requirements are quantified as an input to Technical System Options.

During Logical Design, the Requirements Catalogue is updated as the design 'resolves' the requirements by building them into the design.

A blank Requirements Catalogue entry form is shown in Fig. 5.2.

There are two basic types of requirement — functional and non-functional. The major difference between them is that functional requirements can generally be built into the other SSADM products, and non-functional requirements will define constraints or

Source	Priority	Owner	Requirement ID
Functional requirement(s)			

Non-functional requirement			
Description	Target value	Acceptable range	Comments

Benefits
Comments/suggested solutions
Related documents
Related requirements
Resolution

Figure 5.2 Blank Requirements Catalogue entry form

levels of quality either for a single functional requirement or across a number of functional areas.

Requirements of both types should be quantifiable and unambiguous and should provide a basis for testing the acceptability of the new system.

Functional requirements

Functional requirements include areas such as:

- updates;
- enquiries;
- reports;
- data;
- interfaces with other systems.

This type of requirement is initially documented as part of the fact-finding exercise of Stage 1. Here the users will describe their needs in words, often in general terms. The Requirements Catalogue is an ideal place to document all the requirements in a textual format before they can be unravelled and modelled more precisely by the other SSADM techniques.

Functional requirements are updated as the project proceeds for several reasons:

- Modelling and options highlight the need for a requirement to change.
- Two or more requirements may be shown to be in conflict and need to be resolved.
- Non-functional requirements are added to define the service levels required for a functional requirement.
- The incorporation of the requirement into another SSADM product is indicated.

Non-functional requirements

Non-functional requirements describe how, how well, or to what level of quality a facility or group of facilities should be provided. Therefore, a non-functional requirement will either be attached to a functional requirement or it will be applicable across a wider area of the system. The majority of non-functional requirements will have a target value attached with an acceptable range and therefore provide objective measures of success for the system possibly to be built into criteria for later testing.

There is a wide variety of non-functional requirements. Categories of non-functional requirements include:

- service level requirements;
- access restrictions;
- security;
- monitoring;
- audit and control;
- constraints.

SERVICE LEVEL REQUIREMENTS
Service level requirements are a measure of quality. They define how well the system must perform in a number of aspects, for example:

- hours during which the system will be available and percentage availability required during these hours;

- response times or turnaround times required;
- throughput in terms of number of transactions to be handled or number of records accessed in a period of time;
- reliability of the system as a whole.

ACCESS RESTRICTIONS

Access restrictions may be defined in terms of which data can be accessed by whom, which functions can be accessed by whom, or, at a more general level, the types of security measures to be taken to restrict user access to the system.

SECURITY

Security of the system against unexpected failure can generate requirements in the following areas:

- how often should data be backed up and what mechanism should be used;
- speed and priorities for recovery after failure;
- what facilities are required while recovery is in progress.

MONITORING

Monitoring may be required to check the performance and usage of the system. Requirements for monitoring and reporting can be defined.

AUDIT AND CONTROL

Audit requirements may be for financial, system, or performance reasons. This may require the creation of an 'audit trail' or the production of statistics. Constraints on the entry and checking of data may also need to be defined.

CONSTRAINTS

There are almost always constraints that affect the design of a new system. In many cases, organizational standards dictate certain aspects of the system from the outset. The Requirements Catalogue can be used to document all the constraints that will affect the final design and implementation. Some topics of particular relevance to most systems are:

- conversion of the data from the current system;
- interfaces to other systems;
- human–computer interface issues;
- archiving policy.

Documenting requirements

Figure 5.2 shows an example of a blank Requirements Catalogue entry form. Figure 5.3 shows the same form filled in for a requirement from the Opera Booking System. This will be used to demonstrate the use of the form in SSADM.

The items on the Requirements Catalogue entry are completed in the following way:

- *Source* This is where the requirement came from. This could be a person, as in Fig. 5.3 (Box Office Manager), or a document, etc.
- *Priority* The priority is user-defined. In this case, it is M (medium priority).

Source	Box office manager	Priority	M	Owner		Requirement ID	3

Functional requirement(s)

The new system should provide the box office staff with a method of batching the tickets required for printing so they do not require printing at the time the order is satisfied.

Non-functional requirement

Description	Target value	Acceptable range	Comments
Frequency of printing of tickets	Twice per day	Once per day	Must catch the daily post

Benefits

At present all tickets must be printed at the time the order is satisfied. This means that tickets must stand around awaiting envelopes. This causes tickets to be lost. Batching the production of tickets would help solve this problem.

Comments/suggested solutions

Related documents

Parlimentary report into the efficiency of the opera house

Related requirements

Resolution

Figure 5.3 Requirements Catalogue Entry for Opera Booking System

- *Owner* This indicates the user (or group) responsible for negotiating the requirement.
- *Requirement ID* The unique identifier of this requirement.
- *Functional requirement* A description of the requirement if it is a functional requirement.
- *Non-functional requirement* A definition of the non-functional aspects of a functional requirement or covering a wider area of the system.

- — *Description*   Textual description, possibly reflecting one of the headings in the section on 'non-functional requirements'.
- — *Target value*   This is what should be achieved.
- — *Acceptable range*   This is what would still be acceptable to the user.
- — *Comments*   Any relevant comments that affect the potential resolution of the requirement.
- *Benefits*   Benefits to the user as a result of meeting this requirement.
- *Comments/suggested solutions*   This allows the 'jotting' of ideas in advance of a final resolution of the requirement.
- *Related documents*   Any SSADM or non-SSADM documents may be of relevance here.
- *Related requirements*   A number of requirements may affect the same area of the system. This cross-referencing will assist in the identification of potential conflicts.
- *Resolution*   If this requirement is incorporated into another SSADM document, it should be referenced here. If it is dropped (for example after Business System Options) the reason should be documented here.

5.6   Comparison with Version 3

The technique of requirements definition and the Requirements Catalogue have superseded the Problems/Requirements List of Version 3. The purpose has been expanded to include non-functional requirements and the procedures are more clearly defined. As a result, the Requirements Catalogue has a far more central role in the analysis and specification of requirements than its predecessor, and the Requirements Definition technique has a far higher profile.

6. Function Definition

6.1 Introduction

Function Definition is a procedure that is central to the specification and design of the processing of a system. Functions are the units of processing that users can recognize as being the facilities available to them. Functions can update data within the system or simply enquire upon it. Functions also pull together the various strands of processing definition within other SSADM products to act as a coherent input to physical design and implementation.

Functions are identified, in conjunction with the user, from the Requirements Catalogue, the Required System Data Flow Model, and products of Entity/Event Modelling. They may also be identified as a result of Specification Prototyping. They are cross-referenced to User Roles as an input to dialogue identification (see Chapter 9).

There is no automatic way of identifying functions. This is essentially a creative process. Therefore, the procedures outlined are guidelines rather than 'rules'.

6.2 Naming conventions

The technique of identifying and documenting functions is called *Function Definition*. The documentation of a function is called a *Function Definition*.

I/O Structures are produced in support of each function. These are structured sets of I/O Descriptions showing all the data items input from the user and output to the user for each of the functions. The bottom leaves of the I/O Structure are called *I/O Structure Elements*. The backing documentation for an I/O Structure is an *I/O Structure Description*.

A Function Definition is produced for each of the identified functions which holds both descriptive information about the function (e.g. description and volumes) and references other products in the SSADM set (e.g. processes on the DFDs) to which the function relates.

A Function Definition can contain *service level requirements*, which are used in estimating the size of the system and to define objectives for the system for Physical Design.

A *Universal Function Model* is used to describe the characteristics of every function. It consists of two sets of elements — processes and data streams.

All functions are categorized as follows:

- enquiry or update;
- off-line or on-line;
- user initiated or system initiated.

Each function must belong to one of the options in each of the categories listed above (e.g. a single function may be enquiry, off-line, and user initiated).

6.3 Place in structure

Functions are identified and updated in Step 330: Derive system functions. In addition, Function Definitions are also updated in Step 370: Confirm system objectives, where service level requirements are added.

A number of steps in Stage 3 (Definition of Requirements) will cause updates to Function Definitions. These updates are not implemented in the steps that cause them; instead, changes are fed back into Step 330. This leads to what appear to be circular dependencies. For example Step 330 acts as an input to Step 360: Develop Processing Specification and also references the products from Step 360. In effect the steps are run in parallel and feed directly into one another. The other step with a circular relationship with Function Definition is Step 350: Develop Specification Prototypes.

6.4 Place in Product Breakdown Structure

Function Definitions appear in two places in the Product Breakdown Structure: first in the Requirements Specification and after in the Logical System Specification. This is illustrated in Fig. 6.1. As illustrated, the set of Function Definitions consists of:

- Function Definitions.
- I/O Structures.
- Enquiry Access Paths — for enquiry functions only (see Chapter 11).
- (Common) Elementary Process Descriptions.

This means that the term 'Function Definition' is used in three different contexts:

- the technique name;
- a group name to collect the function details together;
- the description of a function.

Care must be taken not to confuse the different contexts!

6.5 Notation and use

Functions are identified and defined in the Requirements Specification Module. Various components of the functions continue to be defined through the Logical System Specification Module. In Physical Design, Function Definitions and all their associated products are the basis for defining program specifications.

Universal Function Model

Conceptually, all functions can be described in terms of a number of standard components. All functions will have inputs, outputs, and processes. There are two basic elements: processes and data streams. A Universal Function Model can be built up using these two basic elements to describe the characteristics of every function.

A representation of the Universal Function Model is shown in Fig. 6.2. From this model, four different kinds of process can be identified:

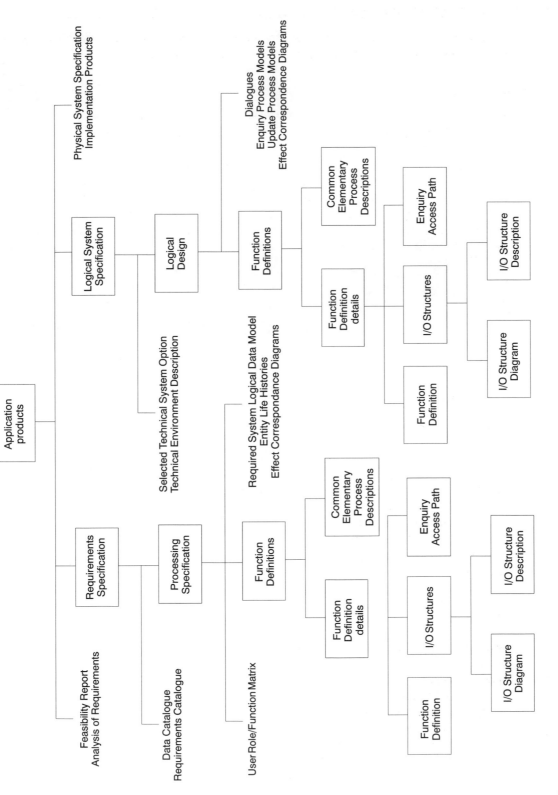

Figure 6.1 Partial Product Breakdown Structure

61

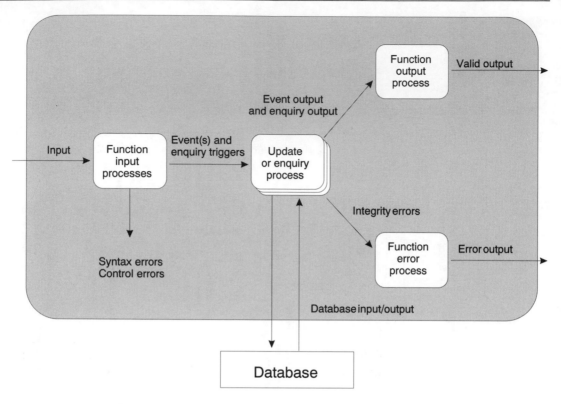

Figure 6.2 Universal Function Model

- Function Input Process.
- Update/Enquiry Process(es).
- Function Output Process.
- Function Error Process.

Also, there are eight different kinds of data streams:

- input (from outside system);
- control/syntax error;
- event/enquiry trigger;
- database input/output;
- event/enquiry output;
- integrity errors;
- valid output (to outside system);
- error output (to outside system).

The principal purpose of this model is to check that all components of each function are defined either before or during Physical Design.

Types of function

All functions are categorized in three ways:

- enquiry or update;
- off-line or on-line;
- user initiated or system initiated.

A combination of these three categories defines the type of a function for subsequent design activities. If a function has any elements that update the database, then the function should be regarded as an update function even if the major part of it is an enquiry.

Deriving functions

Functions are derived from a number of potential sources, depending on whether they are update or enquiry. In general:

- Update functions (and some major enquiry functions) are derived initially from the Required System Data Flow Model.
- Enquiry functions are derived from the Requirements Catalogue or directly in consultation with users.

Events are identified in parallel with the identification of update functions. Events are identified as the triggers to update functions. Subsequent entity–event modelling may identify new events that will, in turn, identify the need for additional functions. Similarly, Specification Prototyping may identify the need for functions to change.

DERIVING FUNCTIONS FROM THE REQUIRED SYSTEM DATA FLOW MODEL

Functions are identified using the bottom-level Data Flow Diagrams as this is where the necessary detail is defined. The derivation of functions from the Required System Data Flow Model is best explained with reference to an example. The level 2 Required System Data Flow Diagram for Process 3 from the Opera Booking System is shown in Fig. 6.3.

User-initiated functions are identified by tracing from an input data flow through the process(es) that use the data to the final outcome of that input (an input/output to a data store or output data flow). In Fig. 6.3, there is an input from the Booking Clerk external entity labelled 'order for tickets' (the Data Flow Diagram also shows that this was originated by a customer). This data flow is used by the process 'receive and validate orders'. Valid orders are then sent to the process 'satisfy orders'. This process updates a number of data stores and sends out details to the Customer and the Accounts section. The presence of a process-to-process data flow indicates that the second process will be taking place within the same time-frame as the first. Therefore, both processes are grouped into the same update function 'receive orders for tickets'.

System-initiated functions are also identified from the Data Flow Diagrams. In this case, processes without an input from an external entity are examined and allocated to functions. In Fig. 6.3, Process 3.3 (print tickets) does not appear to be triggered by an external flow. Instead, it receives input from two data stores and is time-triggered. This corresponds to a system-initiated enquiry function to print out the tickets.

If there are any outputs on the bottom-level Data Flow Diagrams still not allocated to functions, these are traced back to their origin and functions identified for them.

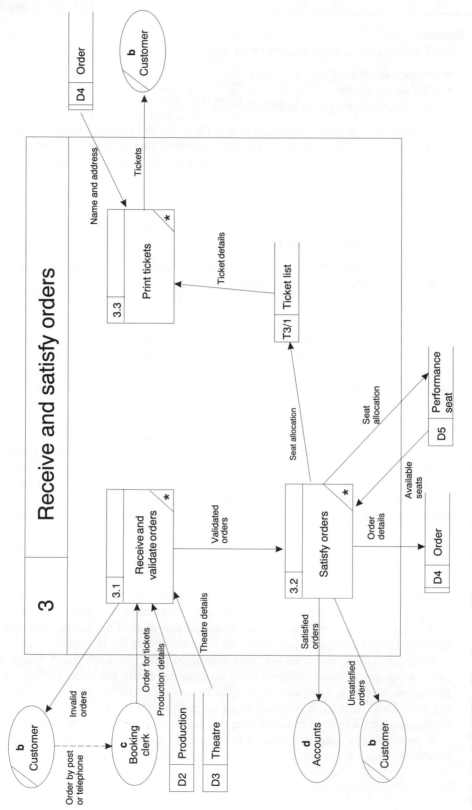

Figure 6.3 Level 2 Data Flow Diagram of Process 3 from Opera Booking System

As demonstrated here, a single function may reference more than one bottom-level process. It is also possible that a single bottom-level process may be referenced by more than one function.

The Requirements Catalogue contains a number of the reports and enquiries identified by the users as being required. These are now documented as enquiry functions. In some cases, the requirement may have been stated in general terms, such as 'Reports are required on all order information'. The precise requirements for reports and enquiries need to be identified for Function Definition, including the data items required by consultation with users.

DERIVING FUNCTIONS FROM OTHER SOURCES

Some of the functions may need to be split into two functions if they are performed both on-line and off-line. Each function must belong to only one of these two types. This does not mean that all the supporting documentation must be duplicated and developed independently — common areas should be developed only once for one function and referenced by the other function.

The initial set of functions identified from the Data Flow Model and Requirements Catalogue are used as an input to entity–event modelling and specification prototyping. In turn, these two activities may identify the need for additional functions or changes to the original functions.

USER CONSULTATION

The identification of functions needs to be checked with the users. It is important that the functions match the needs of the users' work procedures. In some cases, functions may need to be merged if they will always be performed together. Often, an enquiry will always precede an update. In this case, the enquiry should be cross-referenced by the update function.

I/O Structures

An I/O Structure is part of the supporting documentation of a function. It describes the data items that are input to and output from a function. The data items are structured into Jackson-like diagrams showing sequence, selection, and iteration of groupings of data items. These structures are used as the basis of dialogue design in the Logical System Specification Module.

I/O Structures are derived from a number of sources:

- For functions that are derived from the Required System Data Flow Model, the I/O Descriptions are used as a basis for the I/O Structures.
- For functions that are derived from the Requirements Catalogue or as a result of other techniques, lists of required data items need to be derived based on sample reports or directly from the users.

To construct I/O Structures, all the constituent data items input to or output from a function are identified and collected together. Although there may be a number of sources of data items to be considered, there is generally only one I/O Structure for each function (off-line functions may require a separate I/O Structure for the input and output). The I/O Descriptions relevant to the function 'receive orders for tickets' are shown in Fig. 6.4. The data items are grouped into the leaves of a Jackson-like structure using the following guidelines:

- Input and output data items are not included in the same group.
- Repeating data items are not mixed with non-repeating data items.
- Mandatory and optional data items are not mixed. (The mandatory/optional nature of the data item applies within the context of a single function.)

From	To	Data flow name	Data content	Comment
c	3.1	Order for tickets	Customer name Customer address Date of performance Theatre area No. of seats required Price per seat Total for each seat set Total for whole order Method of payment Credit card no. Expiry date	A number of performances can be on one order
3.1	b	Invalid orders	Customer name Customer address Date of performance Theatre area No. of seats required Price per seat Total for each seat set Total for whole order Method of payment Credit card no. Expiry date	The original order is set back with a note detailing the errors
3.2	d	Satisfied orders	Customer name Customer address Date of performance Theatre area No. of seats satisfied	For each of the order lines satisfied
3.2	b	Unsatisfied orders	Customer name Customer address	

Figure 6.4 I/O Descriptions for function 'receive orders for tickets'

These leaves are called I/O Structure Elements. They are structured into a sequence to reflect the sequence of the input and output of data using the basic technique outlined for SSADM Structure Diagrams (see Appendix D). Where I/O Descriptions are used, the Data Flow Diagrams may give an idea of sequence from the context of the data flows. The optionality and iterations are added to reflect the optionality of data being input/output and the repetition of data being input/output. In many cases, the I/O Structure will concatenate I/O Descriptions but it is possible for I/O Descriptions to be interleaved.

The I/O Structure produced in support of the function 'receive orders for tickets' is shown in Fig. 6.5. As may be seen from this diagram, the I/O Structure Elements are annotated to show whether they are input or output.

Each I/O Structure is documented using an I/O Structure Description. This lists the data items contained within each of the I/O Structure Elements. The I/O Structure Description for the function 'receive orders for tickets' is shown in Fig. 6.6. In practice it has been found to be useful to define I/O Structure elements centrally and reference them from a number of different I/O Structures that have similar data input and output requirements.

Documenting functions

Each function is documented on a Function Definition form. The Function Definition form for the function 'receive orders for tickets' is shown in Fig. 6.7. This consists of the following fields (M is mandatory, O is optional):

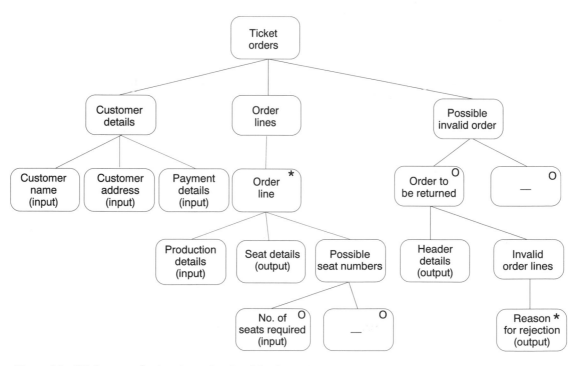

Figure 6.5 I/O Structure for 'receive orders for tickets'

I/O Structure name:	4-1 Ticket orders	
Data flows represented:	c–3.1 3.1–b	
I/O Structure element	**Data item**	**Comments**
Customer name	Customer name	
Customer address	Customer address	
Production details	Date of performance Theatre area	
Seat details	No. of seats Price per seat Total	
No. of seats required	No. of seats	
Header details	Customer name Customer address	
Reason for rejection	Reason for rejection	This will be an internally generated message

Figure 6.6   I/O Structure Description

- *Function name* (M)   A descriptive name that is meaningful to the user.
- *Function ID* (M)   A unique identifier.
- *Type* (M)   Three types must be entered here:
  — update/enquiry;
  — on-line/off-line;
  — user/system (initiated).
  The function in Fig. 6.7 is update, on-line, and user-initiated.
- *User roles* (M for user-initiated functions)   The user roles who will be allowed access to this function and will require dialogue for access.
- *Function description* (M)   A brief description of the function including the circumstances in which it is invoked and any special requirements for presentation to the user.
- *Error handling* (O)   A brief description of the error handling required. This should not include validation checks based on data item definitions in the Data Catalogue but should concentrate on validation needed within the context of the processing of the function.
- *DFD processes* (M for update functions)   The IDs for bottom-level Data Flow Diagram processes related to this function. In this case, there are two processes involved (3.1 and 3.2).

Function name	Receive orders for tickets	Function ID	4

Type	Update/on-line/user-Initiated
User roles	Booking clerk

Function description

This function is invoked to enable the booking clerk to input the details of an order for tickets received from a customer. The system validates the whole order and if necessary rejects it, returning it directly to the customer.

After validation of the order where possible seats are allocated to the order to satisfy it. If the order lines are satisfied then the tickets will be printed (later). If not the customer is informed. All the details of the order are entered into the database.

Error handling

DFD processes	3.1 Receive and validate orders; 3.2 Satisfy orders	
Events 3.1.1 Order receipt	**Event frequency** 1	
I/O Descriptions c–3.1 3.1–b 3.2–d 3.2–b		
I/O Structures 4-1 Ticket orders		
Requirements Catalogue Ref.		
Volumes Average 100 per day; maximum 300 per day		
Related functions 5 Print tickets		
Enquiries None	**Enquiry frequency** N/A	
Common processing None		
Dialogue names Receive order for tickets		

Service level requirements			
Description	Target value	Range	Comments

Figure 6.7 Function Definition form for 'receive orders for tickets'

- *Events* (M for update functions)   For each update function, there will be one or more events to act as a trigger. If there is more than one event for a function, there should be an indication of whether or not the events are mutually exclusive or complementary.
- *Event frequency* (M for update functions by end of Requirements Specification)   The frequency of the event for a single occurrence of the function. In the example above, the event 'order receipt' will occur once for each occurrence of the function. If two events are mutually exclusive, their joint frequency will add up to 1.
- *I/O Descriptions* (M for update functions)   I/O Descriptions for relevant data flows across the boundary for the related Data Flow Diagram processes.
- *I/O Structures* (M)   I/O Structures developed during Function Definition are referenced here.
- *Requirements Catalogue ref* (M for enquiry functions)   Any Requirements Catalogue entry which states the requirement for this function.
- *Volumes* (M by end of Requirements Specification)   Frequency of use of function. Any peaks and troughs in any given time-frame should be indicated. For critical functions, this may reference loading graphs indicating the expected profile of use.
- *Related functions* (O)   Any function that is closely linked with this one. In the example above, the function for printing tickets is related as this function creates the transient data store subsequently used by the 'print tickets' function.
- *Enquiries* (M for functions containing enquiry)   For any function containing an enquiry, there will be one or more enquiries to cross-reference.
- *Enquiry frequency* (M for functions containing enquiry by end of Requirements Specification)   The frequency of the enquiry for a single occurrence of the function.
- *Common processing* (O)   Reference to Elementary Process Descriptions describing common processing that is used by more than one function.
- *Dialogue names* (M for user-initiated functions)   Reference to related dialogue(s) identified at the same time as Function Definition (see Chapter 9).
- *Service level requirements* (M by end of Requirements Specification)   Description, target value, (acceptable) range and comments correspond to the non-functional requirements that may be documented on the Requirements Catalogue Entry form (see Chapter 5).

Other considerations

The presence of a transient data store requires careful consideration in Function Definition. The situation in the example above was fairly clear cut in that the printing of tickets is independent of the receipt of orders. This meant that two functions were identified. However, there may be situations where transient data stores are used to indicate a 'pause' in the middle of two linked processes. In this case, the two processes either side of the transient data store may be included in a single function.

Extra functions may be required for *ad hoc* enquiries. *Ad hoc* enquiries are of two types:

- Enquiries whose output can be fully specified which are run on an *ad hoc* basis.
- Enquiries where the data to be output is only roughly known.

Functions should be identified covering both types of enquiry. Even though the precise nature of the function cannot be specified in detail, the Function Definitions produced will help in the estimation of system sizing and in the selection of facilities required in Technical System Options.

6.6 Comparison with Version 3

Version 3 identifies functions based upon Data Flow Diagram processes. However, they are at a much higher level than the Version 4 functions. The guidelines for the identification of functions in Version 3 starts from the level 1 processes and only drops down to lower levels if a top-level process contains a mixture of batch and on-line processing. By comparison, the Version 4 guidelines start at the bottom-level processes.

In Version 3, update functions are documented in the Function Catalogue and enquiry functions are documented in the Retrievals Catalogue. In Version 4 they are all documented as Function Definitions.

The differentiation of functions into 'types' in each version is slightly different. The types of function identified in each version are:

Version 3	Version 4
Batch updates	Enquiry or update
Online updates	Off-line or on-line
Retrievals	User initiated or system initiated

The cross-referencing of functions to other SSADM products is more clearly defined in Version 4. In Version 3, I/O Descriptions and Dialogues are related to events, not functions. As a result, the role of functions in Version 3 has often been misunderstood and events taken as the 'units of processing'. By comparison, Version 4 relates Dialogues and inputs and outputs directly to functions.

An element of Function Definition that is completely new to Version 4 is the structuring of I/O Descriptions into I/O Structures. Also, the concept of the Universal Function Model has been introduced into Version 4.

7. Options

7.1 Introduction

'Options' are placed at a number of important points within SSADM where enough information has been collected about the system to allow the users formally to select the way forward. The type of option available to the users depends upon the amount and detail of the information collected. At an early stage in the development, the detail will be minimal and therefore the options can only be in broad terms. Later on, when a lot of detail is available, it is possible to be more precise in the description of options available.

Apart from the options in the Feasibility Study Module, which are not covered here, there are two sets of options within SSADM:

- Business System Options (BSO).
- Technical System Options (TSO).

The Business System Options are concerned with broad issues to do with the impact of the system on the organization and the system boundaries. Technical System Options are concerned with more detailed issues concerning the implementation environment needed to support the new system.

Both sets of options have their own stage. The Business System Options stage develops an outline of the new system as an input to the Definition of Requirements. The Technical System Options develop a detailed view of the technical environment as an input to Physical Design. Each of the different types of option are dealt with separately here.

7.2 Business System Options

Business System Options give the analysts and users their first opportunity to break away from their detailed examination and modelling of the current system and consider how best to go about designing a system that will meet the users' requirements. Business System Options allow a complete review of what the system will encompass or whether the development should proceed at all.

Business System Options are a description of what the system is to do without consideration of how it is to do it. In the majority of cases, this is not possible without some broad decisions about features of the technical environment as these will affect the functionality and boundaries of the system. Also in a number of cases some physical characteristics of the new system will be known and can therefore be taken into account at this stage.

A number of Business System Options are developed for consideration by the user. Each option will represent one of a number of possibilities for the new system. In some cases, each option will be a description of the whole system. In other cases, options for discrete areas of the system will be developed.

Business System Options are documented in the most appropriate way, depending upon the nature of the option. Generally, an option will be expressed in terms of a textual description supported by Data Flow Diagrams and a Logical Data Structure if this is considered helpful.

Naming conventions

A number of Business System Options are prepared and one becomes the *Selected Business System Option*. A *cost/benefit analysis* will be produced to accompany each option. An *impact analysis* will also be produced for each option.

Place in the structure

Business System Options are in their own stage within the Requirements Analysis Module coming immediately after Stage 1: Investigation of the current environment.

Stage 2: Business System Options consists of the following steps:

- Step 210: Define Business System Options.
- Step 220: Select Business System Option.

The Selected Business System Option is used as a central input to Stage 3: Definition of Requirements. It is input to the following steps:

- Step 310: Define required system processing.
- Step 320: Develop required data model.

Place in Product Breakdown Structure

The Selected Business System Option is included in the Analysis of Requirements. This is represented in the partial Product Breakdown Structure in Fig. 7.1.

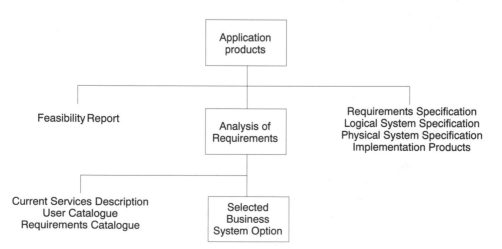

Figure 7.1   Partial Product Breakdown Structure

Notation and use

Business System Options are developed after the current system investigation has been completed. The formulation of the options will be based upon the requirements documented in the Requirements Catalogue. These requirements should indicate to what extent the new system should replicate the functionality and data of the current system. This will determine whether or not the Logical Data Flow Model and Current Environment Logical Data Model are also to be used as input to the options. Other inputs will include the User Catalogue developed in Stage 1, the Feasibility Report (if produced), and the Project Initiation Document.

A number of Business System Options are produced. Normally three separate options are considered sufficient to enable users to make a decision, but if necessary more options can be produced. Each option may include:

- the functional areas to be included within the system;
- inputs and outputs;
- the users of the system;
- external interfaces;
- all requirements addressed by the BSO;
- a partial analysis of the effect on the organization in terms of a cost/benefit analysis and impact analysis.

It is possible that the formulation of Business System Options will start with the development of 'skeleton' Business System Options, a subset of which will be developed further into full Business System Options for subsequent presentation to the users and selection of a single option.

Some of the issues highlighted by Business System Options for the Opera Booking System are as follows:

- Should a record of all customers' names and addresses be kept or only the subscribers?
- Should a record of ticket numbers be kept in case duplicates need to be issued?
- Is a reserve list required when performances are oversubscribed?
- Who is to be allowed access to the system, e.g. does the manager need to use the system?
- Should access to the system be given to ticket agencies?

Many such issues will be based on the requirements of the business and can only be resolved by the users.

7.3 Technical System Options

Technical System Options involve a detailed examination of the functionality of the required system to determine ways in which it can best be implemented. Decisions made at Business System Options are reviewed and a more detailed cost/benefit analysis can be developed to support each option. The aim of Technical System Options is to produce a detailed description of the technical environment to act as an input to Physical Design.

Technical System Options need to take account of constraints on the system, many of which will be documented in the Requirements Catalogue. These will include financial, physical, technical and organizational constraints. The constraints determine how wide-ranging the options can be. For example, if the organization has already made decisions about its strategic development environment, the options concentrate on how the requirements will be supported by the chosen environment rather than looking at alternative environments.

A number of Technical System Options are developed for consideration by the user. Each option will represent one of a number of possibilities for the new system expressed as a Technical Environment Description backed up by a cost/benefit analysis and any other appropriate documentation including a system description, impact analysis and outline development plan.

Technical System Options are undertaken in parallel with the Logical Design of the system.

Naming conventions

A number of Technical System Options are prepared and one becomes the *Selected Technical System Option*. An *Outline Technical Environment Description* is produced to support each option. A full *Technical Environment Description* is developed for the chosen option.

Other products of Technical System Options include:

- application style guide;
- cost/benefit analysis;
- system description;
- impact analysis;
- outline development plan.

Place in the structure

Technical System Options are in their own stage within the Logical System Specification module running in parallel with Stage 5: Logical Design.

Stage 4: Technical System Options consists of the following steps:

- Step 410: Define Technical System Options.
- Step 420: Select Technical System Option.

The Technical Environment Description developed to support the Selected Technical System Option is then used as a central input to Stage 6: Physical Design. It is input to the following steps:

- Step 610: Prepare for Physical Design.
- Step 670: Assemble Physical Design.

Place in Product Breakdown Structure

The Logical System Specification contains the Selected Technical System Option and Technical Environment Description. This is shown in Fig. 7.2.

Notation and use

Technical System Options are formulated after the production of the Requirements Specification. The Requirements Specification contains a detailed description of the functions and data required by the new system, together with their volumes and loading profiles. The Requirements Catalogue and Function Definitions also contain the Service Level Requirements for the system. The requirements and volumes will be used to

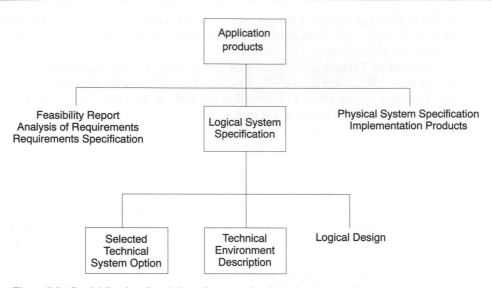

Figure 7.2 Partial Product Breakdown Structure for Technical System Options

determine the size of system required and the features that will be needed to support the users' requirements. The formulation of the Technical System Option involves an exploration of possible technical solutions that will also conform to the selected Business System Option.

Technical System Options are not only constrained by requirements from within the SSADM documentation. Constraints from other sources need to be considered, e.g. cost, timescales, and hardware/software standards (if these have not already been documented into the Requirements Catalogue).

Various technical skills are necessary for Technical System Options which may need specialist assistance:

- *Capacity planning* techniques are required to assess the size of system that will be required to support the functionality and data to the level of service required.
- *Estimating* the time and resource requirements for the remainder of the project is required for the Outline Development Plan.
- *Up-to-date technical knowledge* of the types of development environment available is required to ensure all possible options are explored.
- *Risk assessment and management* skills may be required to ensure that the options do not exceed acceptable levels of risk.

A number of issues may be addressed once a Technical System Option has been selected:

- *Test criteria* can be derived from the Requirements Catalogue, Function Catalogue, and Technical Environment Description.
- *Training* needs for the developers and users of the system can be established.
- *User manuals* can be planned based on the technical environment chosen.
- *Take-on* of current system data can be considered and plans for take on developed.

A number of Technical System Options are developed. It is possible that a number of skeleton options may be refined into a smaller number (probably three) before being presented to the users for selection. Each option is documented in terms of:

- outline Technical Environment Description (TED);
- outline development plan;
- cost/benefit analysis;
- impact analysis;
- system description.

The options are presented to the users and a formal selection takes place. The chosen option is documented more fully before proceeding onto Physical Design.

The Technical Environment Description is a key component of the chosen option. In general it will consist of:

- hardware;
- software;
- system sizing;
- additional items (e.g. fallback and recovery).

The Application Style Guide is developed as part of the Technical System Options. This details standards, based on the Installation Style Guide if one exists, for the development of the human–computer interface. For example, standard layouts for screens and assignment of function keys may be defined. This is also used as an input to Dialogue Design in Stage 5: Logical Design.

For the Opera Booking System, a number of issues are of relevance for Technical System Options:

- The new system will be implemented on existing equipment, upgraded if necessary. This requires a detailed capacity planning exercise to determine the capacity required.
- Printing tickets may require specialist printers. This requires research into printers and a survey of what other opera houses and theatres use for this purpose.
- The numbers and siting of terminals is needed so that the costs of communications hardware and software can be assessed.

7.4 Comparison with Version 3

The purpose of options has not changed very much from Version 3. However, the place and profile of options has changed:

- Business System Options are embedded in Stage 2 in Version 3 but have their own stage in Version 4.
- Technical System Options are before Logical Design in Version 3 but run in parallel with Logical Design in Version 4.

Being at the end of a module means that the Business System Option stage becomes a project decision point where, if necessary, the project can be terminated. This raises the profile and importance of Business System Options from Version 3, where it is simply a precursor to the specification of the required system.

The term Technical Options is used in Version 3 and has been changed to Technical System Options for Version 4.

The guidelines for Technical System Options have been greatly enhanced and more emphasis is placed on the need for external skills in the formulation of options. The Application Style Guide is introduced as a product of Technical System Options in Version 4. Activities that are described as part of Physical Design in Version 3, such as the derivation of test criteria and design of user manuals, are commenced in Technical System Options for Version 4.

8. Entity Life Histories

8.1 Introduction

Entity Life History analysis is the first of two techniques that comprise entity–event modelling. The second technique is Effect Correspondence Diagramming, which is described in Chapter 11.

The purpose of entity–event modelling is to examine in detail the causes of change to the main data of the system, to define the constraints on those changes, and to model their precise effects on elements of the data. Entity Life History analysis examines these aspects from the point of view of the data. Effect Correspondence Diagramming turns this view inside-out by looking at it from the viewpoint of the event and models each cause of change and all the data affected.

Entity Life History analysis is a diagrammatic technique. An Entity Life History is drawn for each entity on the Required System Data Model. It is a pictorial representation of the life of an entity from its creation to its deletion in terms of all the events that may cause a change in the entity and the order in which those events may occur. An event may be thought of as whatever causes data to change, and processes are the system's response to events and the means by which changes are made.

The events that affect entities are initially identified during Function Definition. Entity–event modelling helps to validate the functions in that:

- all events should be incorporated into one or more Function Definition(s);
- functions initially identified as enquiry functions may require reclassification if they are found to affect data;
- the description of the functions may need to be altered in the light of new information being uncovered.

Entity–event modelling also helps to validate the Required System Logical Data Model in that:

- difficulties in drawing Entity Life Histories often indicate that entities are wrongly identified;
- difficulties in drawing Effect Correspondence Diagrams often indicate that new or altered relationships are required;
- new attributes may be required to register the effect of events.

8.2   Naming conventions

Entity–event modelling consists of:

- Entity Life History analysis;
- Effect Correspondence Diagramming.

The Entity Life History analysis technique results in the production of *Entity Life Histories*. Each Entity Life History charts the *effects* of *events* on a single *entity*. An *event* causes a process to update system data. An *effect* is the change within one entity occurrence as a result of a single event.

Operations and *state indicators* are added to the effects on an Entity Life History. An operation is a simple component of processing which describes an action that is performed as part of each effect. As such, one or more operation(s) will make up the processing requirements of the effect. A state indicator may be considered as a 'flag' within the entity that is updated to a new value each time an event affects the entity. Examination of the state indicator value will precisely define the position of an entity in its Entity Life History.

An *Event/Entity Matrix* is constructed as a precursor to Entity Life Histories showing the relationships between entities and events.

The basic elements of an Entity Life History diagram are *sequence*, *selection*, and *iteration*. Additional notations are available for *parallel structures* and *quits and resumes*.

If a single event occurrence has different effects on more than one instance of a single entity, each effect is considered to be acting on a different *entity role*. If an event can affect an entity at several different points in its life, *effect qualifiers* are used to distinguish the different effects.

8.3   Place in structure

Entity Life History analysis is used in two modules of SSADM Version 4, namely Requirements Specification and Logical System Specification. The two steps that are involved are:

- Step 360: Develop processing specification   The Event/Entity Matrix and Entity Life Histories are created. Operations are added to the Entity Life Histories, but state indicators are not.
- Step 520: Define update processes   State indicators are allocated to the effects.

8.4   Place in Product Breakdown Structure

Entity Life Histories appear in the Processing Specification which forms part of the Requirements Specification. This is represented in Fig. 8.1. The Event/Entity Matrix does not appear in the Product Breakdown Structure as it is considered an intermediate product. The Entity Life Histories do not appear in the Logical Design as they are considered to have been superseded by the Effect Correspondence Diagrams and Update Process Models by the end of the Logical Design phase (see Chapters 11 and 12).

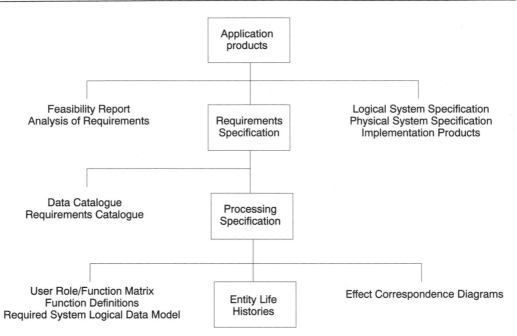

Figure 8.1   Partial Product Breakdown Structure showing Entity Life Histories

8.5   Notation and use

Entity Life Histories are drawn as a tree structure conforming to the basic structuring notation described in Appendix D with the addition of parallel structures, quits and resumes, operations, and state indicators.

The notation consists of:

- sequence;
- selection;
- iteration;
- parallel structures;
- quits and resumes;
- operations;
- state indicators.

These elements are combined into a structure that represents the life of an entity in terms of the events that cause it to be changed. The construction of an Entity Life History combining these elements is demonstrated using an example of the entity Order from the Opera Booking System (see Appendix F) in the following sections.

The Event/Entity Matrix

A very useful starting point for Entity Life Histories is the Event/Entity Matrix. The matrix cross-references the entities from the Required System Logical Data Model to the events that are identified and documented as part of Function Definition (see Chapter

Event	Artist	Artist in performance	Bookable seat	Opera performance	Order	Order line	Part of theatre	Production	Seat at performance
Arrival of order					C	C			M
Details sent to credit card company					M				
Credit card approved					M				
Request for new details sent					M				
New details received					M	M			
Tickets sent					M	D/M			
Performance takes place					M				
Change to personal details					M				
Retention period expires					D	D			
Receive new production details	C	C			M			C	
Performance details finalized	M/D	M	C	C					C
Opera house details changed	M	M	M	M			C/M	M	M

Figure 8.2 Event/Entity Matrix for Opera Booking System

6). As for functions, events are initially identified from the Required System Data Flow Diagrams by determining the triggers to processes that update the main data stores.

An extract from the Event/Entity Matrix for the Opera Booking System is shown in Fig. 8.2. Here, the intersections of the matrix are annotated with 'C' (creation), 'M' (modify), or 'D' (deletion) to denote the type of effect the event will have on the entity. It is possible that the same event has two different types of effect on the same entity. In this case, more than one letter will be put in the intersection.

Before starting the Entity Life History construction, the matrix can be used to check that:

- All entities are created by at least one event.
- Each event affects at least one entity.

It is also worth spending a short time in consultation with the users at this point to check that no obvious events and effects have been missed.

Although this is classified as an intermediate product, practice has shown there to be great practical benefit in maintaining a cross-reference between events and entities from Step 360 through to the end of Logical Design as a reference document. This may also be supplemented by a brief description of each event and the entities affected as this has been found helpful in practice.

The Entity Life History Diagram

Entity Life Histories are usually created in two passes:

- Working from the bottom of the Required System Logical Data Structure to the top, considering the lives of master entities in conjunction with their detail entities.
- Working from the top of the Required System Logical Data Structure to the bottom, considering the lives of detail entities in conjunction with their master entities.

In this way a complete picture is built up of the life of each individual entity and the influences from one life to another.

The types of influence that may be identified in the two passes are:

- the creation or deletion of a detail may change the master;
- the death of the last detail may cause the death of a master;
- the death of the master may cause the death of a detail;
- the death of a master may be constrained not to occur until the death of all the details;
- the death of the master may have no effect on the detail.

Entity Life Histories are started with reference to the Event/Entity Matrix. An entity is selected from the matrix and all events that affect it are listed according to whether the event creates, modifies, or deletes the entity. The events are then organized into a structure representing the life of the entity using the structuring notation identified in Appendix D. Figure 8.3 shows the Entity Life History diagram for the Order entity from the Opera Booking System. The bottom boxes represent the effects of events and should contain the names of events with the possible addition of entity roles or effect qualifiers. Other boxes are used to define the structure.

The Entity Life History in Fig. 8.3 demonstrates some of the different notations available on an Entity Life History.

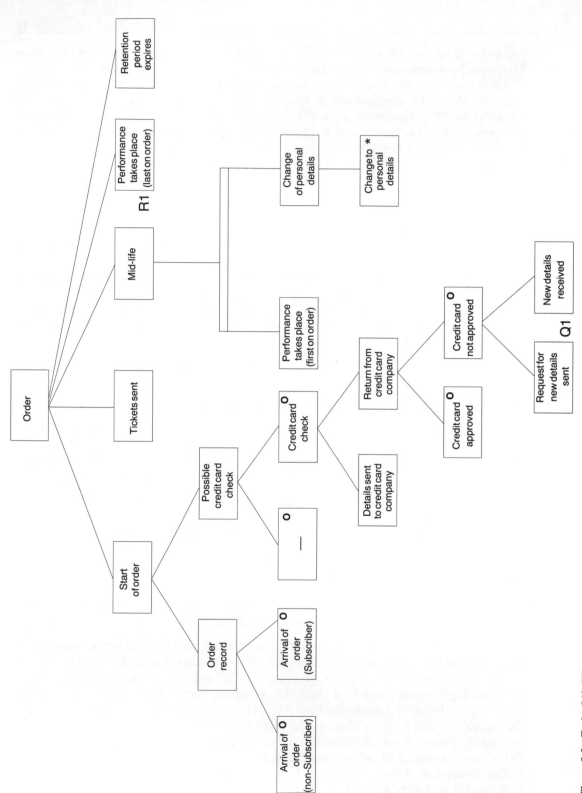

Figure 8.3 Entity Life History for Order

84

SEQUENCE

There is a sequence shown immediately below the top box on the diagram. This shows that the 'start of order' precedes 'tickets sent', which precedes 'mid-life', and so on.

Note that event names can appear at any level in the Entity Life History provided that they do not have boxes below them. Therefore 'tickets sent', which is an event name, appears at the same level as 'mid-life', which is not. Intermediate structure boxes should be named meaningfully to describe in general terms what is happening at that point in the Entity Life History as an aid to reading and checking the ELH.

EFFECT QUALIFIERS

Effect qualifiers are used twice:

- The 'arrival of order' event has two different effects depending on whether the order is from a subscriber or not. Only in the former case will it be necessary to note the subscriber number.
- The 'performance takes place' event has two different effects depending on whether it is the first for this order or the last for this order.

A generally accepted convention is that effect qualifiers are placed in parentheses below the event name. If the same event were to affect different entity roles, the convention is that the role names would be placed in square brackets below the event name.

SELECTION

There is a selection between 'arrival of order (non-subscriber)' and 'arrival of order (subscriber)'. It is denoted by an 'o' in the top right corner of the box. This means that either one or the other will cause the creation of the entity. Similarly, there is a 'null' selection under 'possible credit card check' showing that it will either happen or nothing will happen.

ITERATION

There is an iteration of 'change to personal details'. It is denoted by an asterisk in the top right corner of the box. This means that this box may be repeated a number of times or not at all. Each time it is repeated will be after the completion of the previous one (i.e. they do not overlap).

PARALLEL STRUCTURE

There is a parallel structure with the boxes 'performance takes place (first for order)' and 'change of personal details'. It is denoted by a double bar across the top of the boxes. This means that these two boxes will happen in any sequence. Because there is an iteration below 'change of personal details' this represents the situation where a number of changes may take place to the personal details of the customer and at some point in this period, the first performance itemized on the order will take place. (*Note*: it is necessary to log the first performance in the order as until then the tickets assigned to the order can be exchanged.)

A parallel structure does not imply that events may affect the entity at the same time as one another, it simply states that the sequence is indeterminate. Another point to note is that all boxes under a parallel structure will take place — they are not optional unless specifically denoted as such.

QUITS AND RESUMES

A quit and resume denotes that the sequence of the Entity Life History may be altered if an event occurs out of the normal sequence so that after a 'Qn' the Entity Life History continues at the corresponding 'Rn'. This can either be at a point further on in the Entity Life History or at a previous point.

The quits and resumes on the Entity Life History in Fig. 8.3 show that after a request for new details have been sent to a customer who has failed a credit card check, it is possible that nothing further is heard until all the performances covered by the order have been completed. The event 'performance takes place (last on order)' effectively 'drags' the ELH to the resume point. The point at which the 'Q' is placed shows where the sequence might be changed by the resuming event occurring.

It is important to stress that the presence of a 'Q' does not alter the sequence at all. In the normal situation the effect of 'request for new details sent' is followed by 'new details received', which must be valid (if further invalid details are sent this is not registered by the system). The event that causes the change in the sequence is 'performance takes place' for the final order line on the order. In this way, the quit is conditional on the fact that the event shown by the effect annotated with the 'R' has taken place.

(Note that if 'new details' are received after the first performance, but before the last performance on the order, any order lines for performances that have passed are deleted and the tickets sent for the remaining performances.)

A special case of quit and resume may be used for random events that may affect the entity at any point within a predefined part of the structure (or the whole structure). In this case, the effect box is drawn on the diagram but not linked to any part of the structure. An 'Rn' is placed beside the box. Instead of putting the corresponding 'Qn' after every effect in the relevant section of the structure, a general statement can be added to the diagram which states that it is possible to quit from anywhere within the appropriate part of the structure to the 'Rn' beside the random effect.

This can also apply to an 'off the structure structure', which is a fragment of the Entity Life History that may occur at any point of the main Entity Life History.

Operations

Operations are added to the effects on the Entity Life History during the Requirements Specification Module. Their addition helps to validate the diagram by making it clear what basic processing will be performed for each of the effects. At this point, the description of processing will not be comprehensive as the operations are only defined within the context of a single entity. Further types of operation are identified during logical database process design (see Chapter 12).

Within the context of Entity Life Histories, operations are discrete components of processing which make up each effect. They will generally fall into one of five categories:

- store;
- replace;
- tie;
- cut;
- gain/lose.

STORE

This type of operation applies at creation of the entity which is the first time an attribute, or set of attributes, has a value assigned. Examples of its use are:

- store ⟨attribute⟩
- store keys
- store remaining attributes
- store ⟨attribute⟩ using ⟨expression⟩

REPLACE

This type of operation is used to denote the change in value of an attribute or set of attributes. Examples of its use are:

    replace ⟨attribute⟩
    replace ⟨attribute⟩ using ⟨expression⟩

TIE TO

This operation creates a relationship from a detail entity to its master. It will be used in conjunction with the storing of a value to key or foreign key attributes. It is used in the following way:

    tie to ⟨entity⟩

CUT FROM

This operation deletes a relationship from a detail entity to its master. It will normally be used in conjunction with a change to a foreign key attribute. It is used in the following way:

    cut from ⟨entity⟩

GAIN/LOSE

Gain and lose are operations that maintain a relationship from a master entity to a detail and as such will duplicate the 'tie to' and 'cut from' operations that appear in the detail's Entity Life History. They are not often required and the only reason for including them is if it is important to define precisely when the relationship is created or deleted *from the point of view of the master*. They are used in the following way:

    gain ⟨entity⟩
    lose ⟨entity⟩

TYPES OF OPERATION NOT USED IN ENTITY LIFE HISTORIES

The following types of operations are not normally used in Entity Life History analysis:

- accessing an entity for navigation;
- validation of data;
- error handling;
- manipulating data items prior to writing;
- reading an entity prior to updating.

Operations for creating and deleting the entity need not be included on the Entity Life History, though in practice it has been found sometimes useful to do so. Operations to set and check the state indicator values are not included in the Entity Life Histories.

EXAMPLE OF OPERATIONS ON AN ENTITY LIFE HISTORY

Figure 8.4 shows the Entity Life History for the Order entity with operations added. As shown on the diagram, operations are listed on the diagram with a number. The number is added to the relevant effect(s) in a square box attached to the effect box. As the example shows, it is possible to have the same operation referenced by more than one effect. As all the operations are specific to this entity, none of these operations will appear on any other of the Entity Life Histories. It should be noted that it is possible to have an effect box for which there are no operations as in the two effects of 'performance takes place'. This occurs when the effect results in only a state indicator change with no other updates.

OPERATIONS 'IN PRACTICE'

Although not formally documented as part of SSADM, the following has been found in practice:

• Operations can be customized to fit with the project environment. If the implementation environment is constrained from the outset, it is possible to devise a set of operation types that are relevant to the environment without compromising the Entity Life History analysis technique. The same simple types of operation should be used but the terminology and syntax used may be altered.
• A central operations list helps in the transition to Update Process Models in logical database process design.

The examples show that operation numbers are only unique within the context of a single diagram. This may become unwieldy in later steps during logical database process design where operations are assigned from the event point of view rather than the entity point of view. In this case, it is useful to compile a central operations list for the whole project at an early stage where all operations are assigned a unique number and the individual operation lists are not written onto each of the diagrams. Instead one list is held for the whole system.

State indicators

State indicators are added to Entity Life Histories as the first part of Step 520: Define Update Processing. State indicators do the following:

• Help to validate the structure of an Entity Life History.
• Restate the constraints and sequencing of the Entity Life History in a form suitable for inclusion in the Update Process Models.

A state indicator may be thought of as an attribute of an entity that is updated each time an event affects the entity. Examination of the current state indicator value will describe the position of an entity in its Entity Life History. The values given to a state indicator have no significance provided that each effect assigns a unique value to it. The convention is to start with a value of '1' and increment it by 1 each time an event affects the entity.

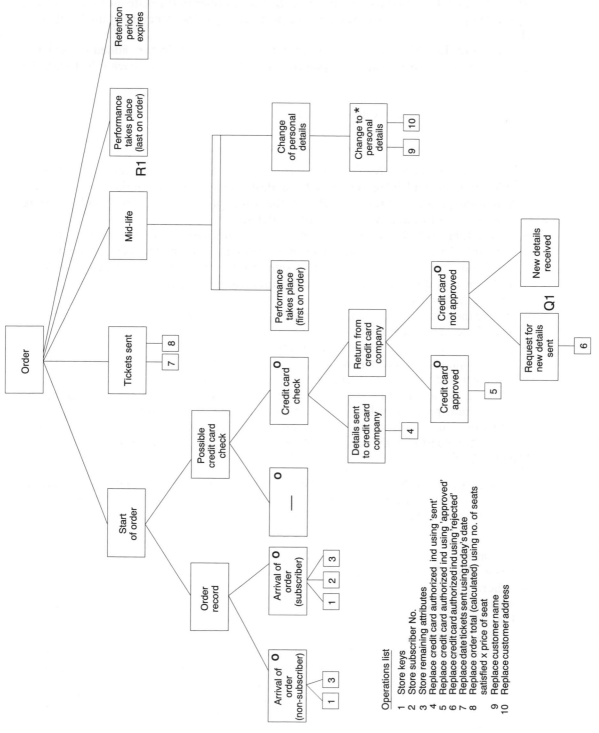

Figure 8.4 Entity Life History for Order entity with operations

Operations list

1 Store keys
2 Store subscriber No.
3 Store remaining attributes
4 Replace credit card authorized ind using 'sent'
5 Replace credit card authorized ind using 'approved'
6 Replace credit card authorized ind using 'rejected'
7 Replace date tickets sent using today's date
8 Replace order total (calculated) using no. of seats satisfied x price of seat
9 Replace customer name
10 Replace customer address

Each effect box on an Entity Life History diagram sets a single unique value of the state indicator. It is also possible to define the set of values of the state indicator that will be valid before the effect is allowed to take place based on the structure of the Entity Life History.

The Entity Life History for the Order entity with state indicators added is shown in Fig. 8.5. State indicator values are placed below the operations for each effect with the following syntax:

⟨valid previous values⟩/⟨value set by the effect⟩.

There are various points to note about the use of state indicators:

- Before an entity is created and after it is deleted, the state indicator value is indeterminate and is represented by a dash. Thus the state indicator values for 'retention period expires' are '9/–' as the valid previous value is 9 and the state indicator is set to null when the entity is deleted.
- Where there is a selection, all effects that are alternatives to one another have the same set of 'valid previous' state indicator values as shown by the fact that 'credit card approved' and 'request for new details sent' both have a valid previous value of 3. (A null selection does not have state indicators assigned as this is a representation of nothing happening.)
- Where there is an iteration, the value set by the repeating effect(s) must also appear as one of the valid previous values.
- An asterisk is used to denote that this effect does not change the state indicator value so it remains at whatever value it was set to before the effect occurred. This is demonstrated by the effect 'change to personal details'.
- Where a quit and resume are used, the value set by the effect with the 'Q' must be one of the valid previous values for the effect with the 'R' in addition to being a valid previous value for the next effect in sequence. This means that the value of 5 is a valid previous value for both 'new details received' and 'performance takes place (last on order)'.

There are two different ways to handle the assigning of state indicator values to parallel structures. Either only one leg updates the state indicator or a separate subsidiary state indicator is introduced for each leg after the first:

- Where only one of the legs of a parallel structure needs to be kept track of, the state indicator is updated by this leg and none of the other legs update the state indicator value at all. This is the situation shown in Fig. 8.5.
- Where all legs of a parallel structure require to be kept track of, an additional state indicator is introduced for each leg other than the first. This is demonstrated in Fig. 8.6.

  This example may be considered as peripheral to the Opera Booking System in that it is logging consignments of scenery into the Opera House. The life of 'consignment' starts with the receipt of the consignment followed by a number of checks that precede its clearance for use in the production. During this period, there may be a number of queries raised by the manager, each of which must be resolved before the next is lodged.

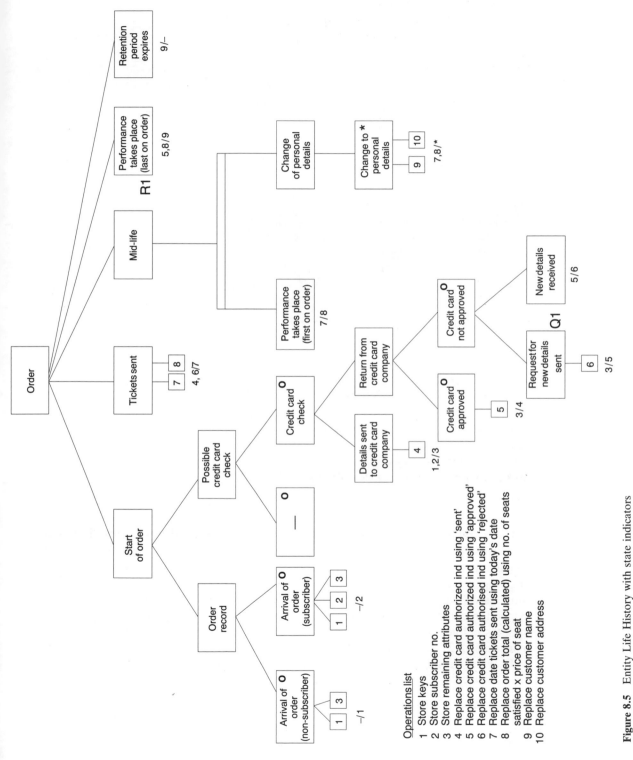

Figure 8.5 Entity Life History with state indicators

Operations list
1 Store keys
2 Store subscriber no.
3 Store remaining attributes
4 Replace credit card authorized ind using 'sent'
5 Replace credit card authorized ind using 'approved'
6 Replace credit card authorised ind using 'rejected'
7 Replace date tickets sent using today's date
8 Replace order total (calculated) using no. of seats
   satisfied x price of seat
9 Replace customer name
10 Replace customer address

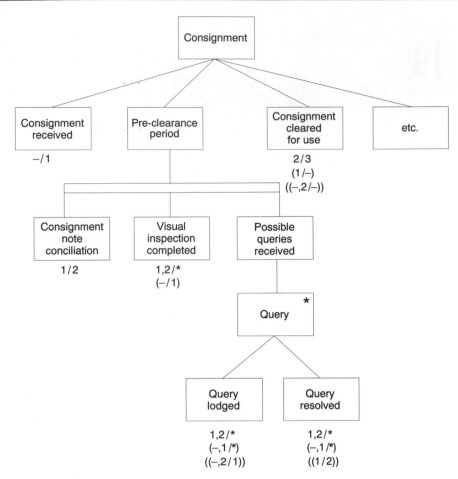

Figure 8.6 Example of subsidiary state indicators

The parallel structure has three legs and therefore needs two subsidiary state indicators in addition to the main state indicator. The effect of 'consignment note conciliation' is to update the main state indicator, but the other two legs leave this unchanged. The reason for showing the first state indicator on the other two legs is to show its valid previous values. The second leg, showing the effect of the event 'visual inspection complete', updates the second state indicator from null to 1. The third leg updates neither of the first two state indicators but updates the third state indicator from null (or 2) to 1 and from 1 to 2. The effect immediately after the parallel structure will set the subsidiary state indicator values back to null.

8.6 Comparison with Version 3

The Entity Life History analysis technique remains basically the same in Version 4 as it was in Version 3 with two significant exceptions:

- Operations are added to the structures to define the effects.
- Quits and resumes have become conditional where they were mandatory in Version 3.

There are a number of more minor changes to the technique, mostly to do with how the interaction of Entity Life Histories is represented.

There is only one set of Entity Life Histories produced in Version 4, whereas there are two in Version 3 — one produced before Relational Data Analysis is performed and the second an updated set based on the new Composite Logical Data Design constructed after Relational Data Analysis.

A terminology change has been introduced in that the ELH Matrix from Version 3 has been renamed the Event/Entity Matrix in Version 4.

9. Dialogue Identification and Design

9.1 Introduction

Dialogue design defines the users' interface with the functions of the new system. Where entity–event modelling is essentially looking at the way in which the processing will interact with the *data*, dialogue design looks at the way in which the processing will interact with the *user*. Dialogue design and entity–event modelling are therefore complementary techniques, which together help to define the functions of the system.

Dialogues are developed in parallel with the definition of the processing of the system. The dialogue view of the system is independent of the processing view and each can then act as a check upon one another at the end of Logical Design. The two views are connected through functions upon which all dialogues are based.

The specification and logical design of dialogues is undertaken using two techniques:

- *Dialogue identification*, which analyses the requirements for dialogues.
- *Dialogue design*, which defines in detail the structure and context of the dialogues.

The identification of dialogues is based upon the user roles that will have access to the functions of the system. Each user role that requires access to a function will probably require some form of dialogue. In some instances, it is possible that different user roles may require different types of dialogue to access the same function. The requirements for dialogue identified here can act as an input to Specification Prototyping.

Dialogue design is based upon the data items input to and output from a function. Each dialogue is based on the I/O Structure that has been developed to support a Function Definition. The dialogue design technique uses the same structuring techniques as those used for I/O Structures enhanced to show the logical phases of a dialogue.

Menus and Command Structures that group the dialogues together and show the valid routes between dialogues are also developed as part of dialogue design.

9.2 Naming conventions

There are two techniques used to define dialogues:

— Dialogue Identification;
— Dialogue Design.

- The *User Catalogue* lists all the on-line users for the system. A *user* is defined as a person who interacts with the system.

- *User Roles* are developed from the User Catalogue and collect together all job-holders who share a proportion of tasks.
- The *User Role/Function Matrix* is developed by cross-referencing the User Roles to the Functions.
- A *dialogue* is defined initially as the cross-reference of a User Role with a Function and after rationalization may indicate a number of user roles cross-referred to the same function.
- A *Dialogue Structure* is developed for each of the Dialogues showing how the different parts of the Dialogue fit together. Each Dialogue Structure consists of a number of *Dialogue Elements* each of which represents a collection of data items. Each Dialogue Element is documented in terms of a *Dialogue Element Description*.
- *Logical Groupings of Dialogue Elements* (LGDEs) are defined as a collection of Dialogue Elements. The different paths through a dialogue are defined in terms of the LGDEs in a *Dialogue Control Table*. Any 'help' appropriate to the dialogue is defined within a *Dialogue Level Help*.
- *Menu Structures* define a hierarchical representation of the way in which the dialogues fit together, while *Command Structures* define where control goes next on completion of the dialogue.
- A *style guide* defines the standards to be applied to the design of dialogues. There are two of these — the *Installation Style Guide*, defining the standards to be applied across all projects; and the *Application Style Guide*, defining the standards to be applied on the specific project.

9.3   Place in structure

Dialogue identification is undertaken in the Requirements Analysis and Requirements Specification Modules. Dialogue design is undertaken during the Logical System Specification Module. The steps involved are:

- Step 120: Investigate and define requirements
  The User Catalogue is developed.
- Step 310: Define required system processing
  The User Roles are identified.
- Step 330: Define system functions
  The User Catalogue and User Roles are checked for completion. The User Role/ Function Matrix is developed, identifying the dialogues required and which of the dialogues are critical.
- Step 510: Define user dialogues
  The Dialogue Structures are produced and the Logical Groupings of Dialogue Elements are identified. The dialogue navigation, Menu and Command Structures, and Dialogue Level Help are defined.

9.4   Place in Product Breakdown Structure

Figure 9.1 shows how the various elements associated with dialogues are reflected in the Product Breakdown Structure. The User Catalogue is part of the Analysis of Requirements, the User/Role Function Matrix is in the Processing Specification, which is part of

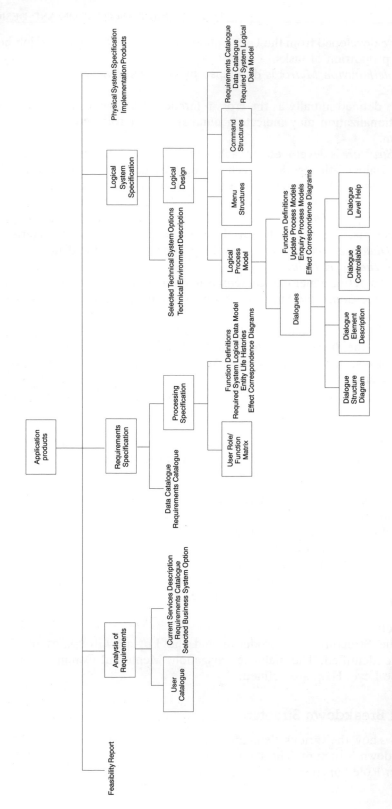

Figure 9.1 Partial Product Breakdown Structure

the Requirements Specification, and the definitions of the dialogues are part of the Logical Process Model, which is part of the Logical Design, as are the Menu Structures and the Command Structures. The Logical Design is itself part of the Logical System Specification.

9.5 Notation and use

Dialogue identification starts with an examination of the users of the system. A User Catalogue is initiated at an early stage, detailing who the users are and the tasks they perform. From this, User Roles, which are groupings of users based on their functions and responsibilities, are identified. The User Roles are cross-referenced to the functions in order to find out what dialogues will be needed. The dialogues identified may form the basis of Specification Prototyping.

Dialogue design develops Dialogue Structures based on the I/O Structures that have been developed in support of functions. The structures indicate how groups of data items are passed between users and the system. These structures are enhanced to indicate logical groupings that may form the basis of future screen design. Menus and Command Structures define the navigation between dialogues, and Dialogue Control Tables define the navigation within dialogues.

Dialogue design should be carried out with reference to a style guide, which lays down the standards to be used in all dialogues. This may be a guide for a whole organization or a guide developed for this specific application. The Application Style Guide is optionally developed as part of Technical System Options.

The techniques and notation are described below.

Dialogue identification

The identification of requirements for dialogues is a thread running through the Requirements Analysis and Requirements Specification Modules. It is achieved by a number of tasks within these modules:

- production of User Catalogue;
- identification of User Roles;
- identification of dialogues required;
- identification of critical dialogues.

None of these involves a diagrammatic notation. Instead, for each task, a form or matrix is produced. The identification of dialogues is demonstrated in the following sections with examples from the Opera Booking System (see Appendix F).

Production of User Catalogue

The User Catalogue lists all the individual users of the system and describes each of their tasks within the context of the system. The catalogue is compiled during the investigation of the current environment, which could encourage the cataloguing of the current users only. However, it is important to anticipate the identity of the users of the new system as well. This means that it may not be possible to complete the User Catalogue until after the selection of Business System Options as one of the areas that will be discussed during these options is the users of the new system.

Job title	Job activities description
House day manager	Organize the performance details
	Set up and amend artist involvement
Opera house manager	Input and amend production details
	Organize the performance details
	Set up and amend artist involvement
Booking office clerk	Receive and action personal booking applications
	Receive and action telephone booking applications
Postal office clerk	Receive and action postal booking applications

Figure 9.2 User Catalogue for the Opera Booking System

The User Catalogue for the Opera Booking System is shown in Fig. 9.2. In the Catalogue, job titles are listed rather than individuals. This is because personnel changes should not affect the responsibilities of a particular post. Some organizations may already have job descriptions for their staff, which would be a useful starting point for the User Catalogue.

Identification of User Roles

A User Role is geared to the usage of the new system rather than any specific job responsibilities. Each User Role defines an area of work or responsibility that will need to use the new system for a particular purpose or will share the same access permissions and restrictions. Each User Role will encompass a number of users and, potentially, a single user could adopt more than one User Role in different situations.

The User Roles are derived from the User Catalogue by examining the tasks performed. If several users perform the same tasks, they will be candidates for being combined into the same User Role. The external entities on the Required System Data Flow Diagrams may also be a useful pointer to User Roles and vice versa. The objective is to group together users who need to access the same functions and data. As the identification of User Roles is very much associated with the way in which the business will function, the identification of User Roles must be done in consultation with the users. If a User Role is identified that is not meaningful to the users, it should be redefined to be recognizable in the business environment.

The User Roles for the Opera Booking System are shown in Fig. 9.3. Here, a number of job titles have been combined as the job tasks are similar and they have similar responsibilities.

User role	Job title	Activities
House manager	House day manager	Organize the performance details
	Opera house manager	Set up and amend artist involvement
		Input and amend production details
Booking clerk	Booking office clerk	Receive and action personal applications
	Postal office clerk	Receive and action telephone applications
		Receive and action postal applications
		Dispatch tickets to customers
Information clerk	Information clerk	Answer telephone enquiries on performances
		Answer telephone enquiries on applications

Figure 9.3   User Roles for the Opera Booking System

In identifying User Roles, consideration should not only be given to which functions will be used by groups of users but also the way in which the functions will be used. For example, some groups of users may have frequent access to the system and others may have infrequent access to the system. The latter group may require a more interactive type of dialogue to lead them through a particular function and provide more assistance than the former group, who will quickly become familiar with the system and require less help each time.

Although not a primary consideration in the identification of User Roles, experience has shown that access security to functions and data can help to determine different User Roles. Many systems restrict access to data but not functions. This would lead to a situation where users may access all functions but access to data will be constrained. If different groups of users will have different levels of restriction applied, they should be separated out into different User Roles. In this situation, User Roles have a wider use in a project than simply defining requirements for dialogue.

Identify the dialogues required

Dialogues are identified as part of the activity of Function Definition. Dialogues are based on functions in that each function with an on-line element will require one or more dialogues.

The identification of required dialogues is assisted by the use of a User Role/Function Matrix. The User Role/Function Matrix is developed after the functions have been identified and documented. It cross-references the User Roles to the functions and shows which functions will be used by which User Roles.

Figure 9.4 User Role/Function Matrix for Opera Booking System

Function / User role	Receive orders for tickets	Print performance advance details	Maintain production details	Maintain performance details	Maintain theatre details	Display application details	Display production/performance details	Produce application statistics
House manager	X	X	X	X	X	X	X	X
Booking clerk	X					X	X	X
Information clerk		X				X	X	

Figure 9.5 User Role/Function Matrix showing critical dialogues

Function / User role	Receive orders for tickets	Print performance advance details	Maintain production details	Maintain performance details	Maintain theatre details	Display application details	Display production/performance details	Produce application statistics
House manager	X [1]	X [1]	X	C	X	X [1]	X [2]	X [1]
Booking clerk	C [2]					X [1]	C [1]	X [1]
Information clerk		X [1]				C [1]	C [1]	

Wherever a function is to be accessed on line by a User Role, an 'X' is placed in the intersection of the matrix. This is demonstrated in Fig. 9.4 for the Opera Booking System. As can be seen from this matrix, the House Manager will want to access all functions, whereas the Booking and Information Clerks will only be interested in a subset.

Each 'X' identifies the need for a dialogue. In Fig. 9.4, the function 'display application details' will be accessed by the three different User Roles. Potentially, each User Role could need a different type of dialogue to access this function. (This does not imply that different data will be input or output but there may be a different style of presentation required depending upon the User Role.) For each function, therefore, each of the relevant User Roles should be considered in terms of their need for the presentation of dialogues and the number of different dialogue types determined. In most cases, it is likely that only one dialogue will be required for each function to be used by all the User Roles able to access the function.

There is no formal way of documenting the dialogues rationalized from the matrix. A possible way of doing this would be to annotate the intersections of the matrix with a number or identifier to show which User Roles will be using a single dialogue. Therefore, if all the intersections belonging to a single function are annotated with a '1', this would mean that they are all to be combined into a single dialogue. This suggested notation is demonstrated in Fig. 9.5. It should be stressed that this is not a formal part of the notation and the identified dialogues could be documented in a variety of different ways. It can be seen that a single dialogue would be produced for the function 'print performance advance details', whereas two separate dialogues are produced for the function 'receive orders for tickets'.

The construction of the matrix and identification of dialogues needs to match the users' perception of the way the system will work, so it is especially important that the users are consulted at all points.

The construction of the User Role/Function Matrix helps to provide a check on the work done to date. Each function can be checked to see if all User Roles are listed that are known to require access and the User Roles can be checked to ensure they have functions to perform all their tasks. This checking may give rise to additional User Roles and functions being identified.

Identification of critical dialogues

Once the requirements for dialogues have been identified, it is possible to evaluate the dialogues in terms of their importance to the system. The most important dialogues can be denoted 'critical'. A critical dialogue is one that is either of high volume or is of most value to the users of the system. The high volume dialogues will influence capacity requirements and affect performance and the 'valuable' dialogues will need to have most attention in design. The identification of critical dialogues will therefore be a useful input to:

- Technical System Options in capacity and performance calculations;
- Specification Prototyping to identify which dialogues should be prototyped.

Factors that will determine whether a dialogue is critical or not include:

- high frequency of use;
- complexity of data access and manipulation;

- importance to the business of the system;
- political sensitivity;
- novelty to the business or users.

In some systems, it may be considered that all dialogues are critical but in the majority of systems there will be a core set of dialogues that can be identified as being more critical than others. For example, the dialogues to maintain non-volatile look-up tables will generally be less critical than the central enquiries of the system.

Critical dialogues can be denoted on the User Role/Function Matrix as shown in Fig. 9.5. Critical dialogues are denoted with a 'C' instead of an 'X'. As may be seen from this matrix, it is possible that some dialogues are critical to only one of the User Roles that use it. For example, there will be only one dialogue for the function 'display application details' and this will be critical to the Information Clerk but not the House Manager or Booking Clerk. If Specification Prototyping is used for this dialogue, this matrix shows that it is more important to consult about and agree the dialogue with the Information Clerks than the other two User Roles.

Dialogue design

The logical design of dialogues is undertaken in parallel with the Logical Database Process Design of the update and enquiry functions as part of the Logical System Specification Module. The dialogue design technique is based upon the I/O Structures identified as part of Function Definition and the User Role/Function Matrix rationalized to identify the dialogues required.

The tasks of dialogue design are:

- production of the Dialogue Structures;
- identification of Logical Groupings of Dialogue Elements;
- identification of dialogue navigation;
- design of Menu Structures and Command Structures;
- definition of Dialogue Level Help.

The tasks of dialogue design are described in the following sections.

PRODUCTION OF DIALOGUE STRUCTURES

The I/O Structure for a function shows the input and output requirements for the function and therefore is used as the basis for the required Dialogue Structure. In fact, the Dialogue Structuring technique simply extends and enhances the I/O Structures rather than producing a completely different set of products. As described in Chapter 6, I/O Structures are Jackson-like structures that conform to the general structuring rules described in Appendix D. Dialogue Structures conform to the same rules with some additional enhancements described in the following sections.

The initial Dialogue Structure for the function 'receive orders for tickets' is shown in Fig. 9.6. This is a direct reflection of the I/O Structure for the function.

The bottom boxes on Dialogue Structures represent groups of data items that are input to or output from the system. They are called 'dialogue elements'. Each dialogue element is either 'input' or 'output' and this is annotated on the diagram as shown above. Dialogue elements are documented further using 'dialogue element descriptions'. At this stage the cross-reference between the dialogue element name and the data items that

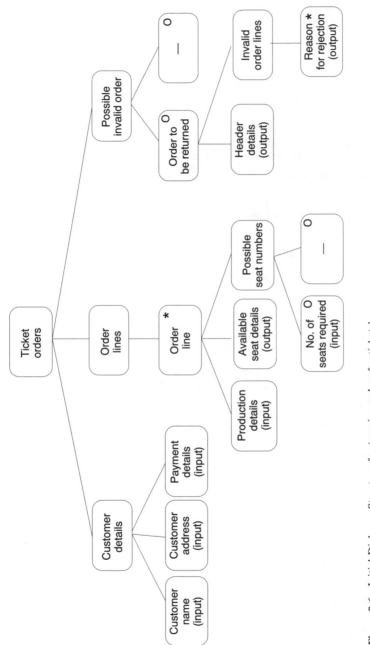

Figure 9.6 Initial Dialogue Structure for 'receive orders for tickets'

Dialogue element	Data items	ID of LGDE	LGDE mandatory or optional
Customer name	Customer name Subscriber no.		
Customer address	Customer address		
Payment details	Method of payment		
Production details	Title of opera Part of theatre Price per seat		
Available seat details	Available seats for requested area Available seats for other areas		
No. of seats required	No. of seats		
Header details	Customer name Customer address System date		
Reason for rejection	Title of opera Date of performance Reason for rejection		

Figure 9.7   Initial Dialogue Element Description for 'receive orders for tickets'

make up the dialogue element are listed on the Dialogue Element Description form. The initial Dialogue Element Description for 'receive orders for tickets' is shown in Fig. 9.7. This is now enhanced further.

IDENTIFICATION OF LOGICAL GROUPINGS OF DIALOGUE ELEMENTS (LGDEs)
The Dialogue Structure is now developed further to reflect logical phases of the dialogue based upon the users' perception of how the dialogue will work in practice. This involves identifying LGDEs which are logically related and will probably be implemented together. Conceptually, a logical grouping of dialogue elements is a logical 'screen' and this perception might help in the identification of groupings. However, these are *not* physical screens and therefore physical considerations (such as the number of characters that will fit onto a screen) should not be used when identifying the groupings.

The following are some of the considerations that apply when identifying logical groupings of dialogue elements:

- A logical group is in the form of a 'message pair' when an input (or a number of inputs) is grouped with a corresponding output (or a number of outputs).
- Different types of structure element (i.e. sequence, selection, iteration) should not normally be mixed within one grouping.
- Groupings may reflect logical phases of a user's task.

Another consideration when identifying Logical Groupings of Dialogue Elements is that the end of each LGDE may represent a point at which it is possible to 'abort' the dialogue.

The logical groupings of dialogue elements are annotated on the Dialogue Structure as shown in Fig. 9.8. As can be seen, each of the logical groupings is uniquely identified on the diagram and these identifiers are added to the Dialogue Element Descriptions as shown in Fig. 9.9.

IDENTIFICATION OF DIALOGUE NAVIGATION

In practice, not all elements of a dialogue are used each time the dialogue is invoked. For example, if the structure shows an iteration, the iterated part may, sometimes, not be performed as the iteration normally includes the 'zero' option. Similarly, where there is a selection including a 'null' box, this part may not be performed.

It is useful to gain an understanding of the possible navigation paths through the dialogues for several reasons:

- Some paths through the dialogue may be used the vast majority of the time. In this case, the dialogue design should be optimized around the more commonly used paths.
- In assessing volumetrics, the dialogue navigation paths will give a more accurate estimate of loadings.

A Dialogue Control Table can be produced for each dialogue showing the different possible navigation routes through the dialogue together with an assessment of their percentage usage. All the paths are defined in terms of the Logical Groupings of Dialogue Elements. The default path is identified initially and all other possible paths are considered.

As an input to constructing a Dialogue Control Table, the Dialogue Element Descriptions are updated to show which Logical Groupings of Dialogue Elements are to be mandatory every time the dialogue is invoked and which are optional. The updated Dialogue Element Description for 'receive orders for tickets' is shown in Fig. 9.10. Here, it can be seen that it is mandatory that the customer details are entered and order lines processed, but that the seat requirements input and returned order output are optional. Note that the input of production details is mandatory even though it is under an iteration. The iteration must therefore be one-based rather than zero-based.

Each path through the dialogue will need to include all of the logical groupings of dialogue elements denoted as mandatory and may include a combination of groupings denoted as optional.

The Dialogue Control Table for 'receive orders for tickets' is shown in Fig. 9.11. The occurrences indicate the number of times the logical grouping of dialogue elements will be used each time the dialogue is invoked. For groupings of dialogue elements in a sequence, this will always be '1'. For those groupings under an iteration, it is useful to estimate roughly how many times they will be used per dialogue usage as an input to sizing.

Note that 'REC-ORD-2' is shown to have a minimum occurrence of 1, even though it is under an iteration. In this way it is possible to distinguish between zero-based iterations and one-based iterations.

An 'X' is used to show which groupings will be used for each path. Any number of paths may be defined such that the percentage path usage adds up to 100 per cent.

106

Figure 9.8 Dialogue Structure with LGDEs

Dialogue element	Data items	ID of LGDE	LGDE mandatory or optional
Customer name	Customer name Subscriber no.	REC-ORD-1	
Customer address	Customer address		
Payment details	Method of payment Credit card no. Credit card expiry date		
Production details	Title of opera Part of theatre Price per seat	REC-ORD-2	
Available seat details	Available seats for requested area Available seats for other area		
No. of seats required	No. of seats	REC-ORD-3	
Header details	Customer name Customer address Subscriber number System date	REC-ORD-4	
Reason for rejection	Title of opera Date of performance Reason for rejection	REC-ORD-5	

Figure 9.9   Dialogue Element Descriptions with LGDEs added

However if the dialogue contains a great number of paths it is possible to show only the major ones.

DESIGN OF MENU STRUCTURES AND COMMAND STRUCTURES

As well as defining the structures of the individual dialogues, it is necessary to define the context of each dialogue within the whole application for each User Role. This can be done in one of two ways: *Menu Structures* or *Command Structures*.

Menu Structures are used where the intention is to implement the system using conventional menu hierarchies. Command Structures are used where it is intended that direct navigation from one dialogue to another will be possible. Often, systems are implemented using a mixture of the two approaches. In this case, both Menu and Command structures may be specified.

Menu Structures

A Menu Structure can be developed for each User Role. The User Role/Function Matrix is used to identify all the functions that can be initiated by a single User Role. From this it

Dialogue element	Data items	ID of LGDE	LGDE mandatory or optional
Customer name	Customer name Subscriber no.	REC-ORD-1	Mandatory
Customer address	Customer address		
Payment details	Method of payment Credit card no. Credit card expiry date		
Production details	Title of opera Part of theatre Price per seat	REC-ORD-2	Mandatory
Available seat details	Available seats for requested area Available seats for other areas		
No. of seats required	No. of seats	REC-ORD-3	Optional
Header details	Customer name Customer address Subscriber no. System date	REC-ORD-4	Optional
Reason for rejection	Title of opera Date of performance Reason for rejection	REC-ORD-5	Optional

Figure 9.10   Mandatory/optional indication added to Dialogue Element Description

ID of LGDEs	Occurrences			Default and alternative pathways		
	Min.	Max.	Av.	Default	Alt 1	Alt 2
REC-ORD-1	1	1	1	X	X	X
REC-ORD-2	1	4	2	X	X	X
REC-ORD-3	0	4	2	X	X	
REC-ORD-4	0	1	1		X	X
REC-ORD-5	0	4	2		X	X
Percentage path usage				85	12	3

Figure 9.11   Dialogue Control Table for 'receive orders for tickets'

should be possible to identify groups of related functions. Like tasks should be grouped together into the same menu area. It is possible that the Data Flow Model hierarchy will suggest groups. However, this should be checked with the users to ensure that these groups match their perception of what logically goes together. The dialogues will be at the bottom level of the menu hierarchy.

The Menu Structure for the House Manager User Role is shown in Fig. 9.12. The structure is a simple hierarchy with soft boxes representing the dialogues (or functions) and hard boxes representing the nodes of the hierarchy. Each node represents a menu and structure nodes and dialogues may both be at the same level. Also, there is no reason why the same dialogue cannot appear in several different places in the hierarchy.

Command Structures

A Command Structure documents the options available to the user on termination of a dialogue. In a simple menu structure, it is usually assumed that on termination of a dialogue, control will pass back up the hierarchy to the node above or the top node. A Command Structure allows control to be passed directly to another dialogue or a menu node, providing a 'fast path' through the system. This can be in addition to a Menu Structure or instead of a Menu Structure.

A Command Structure details the options for a single dialogue. This is demonstrated in Fig. 9.13. In this case, the Command Structure is used in conjunction with the Menu Structure to denote either that control will pass to the main menu or that 'sideways jumping' is allowed to the dialogues to display application details and produce application statistics. The other option here is to repeat the same dialogue again.

It should be noted that all entries within the Command Structure must be valid for the User Role using the dialogue.

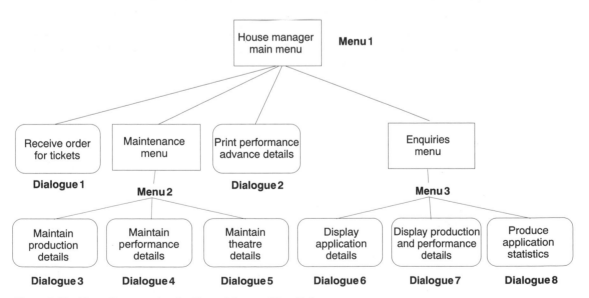

Figure 9.12 Menu Structure for the House Manager User Role

Option	Dialogue or menu	Dialogue/menu name
Register more applications	Dialogue	Receive orders for tickets
Quit to main menu	Menu	Office manager main menu
Produce application statistics	Dialogue	Produce application statistics
Display application details	Dialogue	Display application details

Figure 9.13 Command Structure for 'receive order for tickets'

DEFINITION OF DIALOGUE LEVEL HELP

Most systems will require help facilities to be added to dialogues and menus to provide assistance to users that are unfamiliar with the system. SSADM defines help to be in three categories:

- contextual;
- job related;
- navigational.

Help is an issue that may be addressed during logical or physical design of the dialogues. Any 'help text' identified for the dialogue should be documented textually with it.

STYLE GUIDE

A style guide is a set of standards to define general parameters of screens, dialogues and reports. It is an important guide to developing dialogues as it helps to ensure that the implemented system will have the same 'feel' throughout. For example, a style guide may define how function keys will be assigned so that a certain function key will always do the same thing whatever the context (e.g. F1 = Help). Some of the areas addressed by such a guide will be purely presentational and others will concern the way in which the human–computer interface will operate.

SSADM defines two style guides:

- Installation Style Guide;
- Application Style Guide.

The Installation Style Guide is assumed to cover standards for an entire organization and defines general principles of dialogue design. The Application Style Guide is produced for a specific project and will translate the general into the specific.

The style guides are referenced during dialogue design to ensure that the logical definition conforms with the standards in the guide.

Style guides can be very detailed and time consuming to develop from scratch. There are some commercially available style guides that can be tailored for use by an organization or project. In the absence of any style guide it is important to address standards for dialogues before implementation to facilitate user acceptance of the system.

9.6 Comparison with Version 3

Logical Dialogue Design in Version 3 has been totally redeveloped for Version 4. In Version 3, dialogues are based on events and use a flow-charting type of notation. In Version 4, dialogues are based on functions and use a Jackson-like representation.

In Version 3, dialogues are more intimately connected to the logical process design. In Version 4, within the logical design, the dialogues are developed independently of the process models.

10. Specification Prototyping

10.1 Introduction

Prototypes are used to model aspects of the required system in advance of full development for several reasons:

- They aid communication between developers and users.
- They enable users to experience aspects of the system before implementation.
- They support the formulation of requirements for systems.

The use of prototyping in systems development is becoming increasingly popular as it is seen to reduce the risks of system development by 'trialling' aspects of the system before the full costs of implementation are committed.

There are two broad categories of prototypes:

- those that are designed to be 'thrown away';
- those that are designed to be developed into the final system.

The 'throw-away' prototypes are generally used to help in understanding the requirements for the system, and the development prototypes are used for rapid or incremental development of a system where the requirements are well understood.

Although all types of prototyping can be used in conjunction with SSADM, the type of prototyping that is most compatible with its structured approach is 'specification prototyping', which produces 'throw-away' prototypes. Specification prototyping is incorporated into the Requirements Specification Module and is used to demonstrate selected parts of the Requirements Specification which cover menus, screens and reports.

The prime objective of specification prototyping is to help the user understand what is being proposed by the developers so that they can comment on what is actually required. Its main focus is on helping to:

- identify, clarify, and verify the user's business functional requirements;
- define and verify system characteristics for menus, screens and reports;
- evaluate the appropriateness and potential impact of new design areas;
- give users an initial 'feel' for the system facilities.

Prototypes may be developed using the target implementation environment if this is known from the outset. However, it is possible — and sometimes preferable — to build specification prototypes using different hardware and software from the target environment.

Specification prototyping is often iterative as the requirements may need to be refined a number of times based on user feedback. In this case, the prototyping exercise needs to be carefully controlled to avoid the situation of 'never-ending' prototyping. The objective should be clear from the outset:

- it *is not* to build a perfect demonstration of exactly how the system will look;
- it *is* to help clarify the requirements which will be input to logical and physical design where precise requirements will be addressed.

Specification prototyping should not be considered as a 'mandatory' step within SSADM. It is most appropriate where the requirements are complex and unclear or where users are unfamiliar with the use of computer systems. It helps to reduce greatly the risk of incomplete or misunderstood requirements for a system.

10.2 Naming conventions

Specification prototyping is the only type of prototyping included within 'core' SSADM. The documentation products produced during the prototype activity are as follows:

- *Prototype Pathways*, which are produced for each of the individual prototypes to detail the menus, screens and reports that are to be demonstrated.
- *Prototype Demonstration Objective Documents*, which are produced for each of the individual pathways to assist in defining the object of, and the way in which, the demonstration will take place.
- *Prototype Result Logs*, which are produced for each of the individual pathways to document the result of the demonstration and to define any follow-up action required. A Prototype Result Log is produced for each of the individual pathways to document the result of the demonstration and to define any follow-up action required.
- *Menu and Command Structures*, which are used as the basis for those produced during Dialogue Design in the Logical System Specification Module.

10.3 Place in structure

All the main activities of specification prototyping take place in Step 350: Develop specification prototypes.

10.4 Place in Product Breakdown Structure

Specification prototyping is considered to be a support technique to the main techniques and procedures of SSADM. As such, none of the products of specification prototyping is considered to be a 'deliverable' of the project. Instead, the output of the prototyping exercise is used to enhance other SSADM documentation. The Menus and Command Structures, even though they are used in a later module, are not considered to be formal products but working documentation that will be incorporated into the products of dialogue design. This means that specification prototyping products do not appear in the Product Breakdown Structure.

10.5 Notation and use

Specification prototyping is undertaken in parallel with Function Definition and entity–event modelling during the definition of requirements. The prototypes are based on the functions identified in Step 330 and the User Role/Function Matrix. They are used to confirm the definitions of functions and events as well as providing an input to the Requirements Catalogue. If Menus and Command Structures are prototyped, these are used as the basis for Menus and Command Structures developed as part of dialogue design in the Logical System Specification Module. The results of prototyping may also affect the Logical Data Model and Data Catalogue if data definitions are found to be incorrect or incomplete.

Specification prototyping is undertaken in eight activities:

- Define the Prototype Scope.
- Prototype the Menu and Command Structures.
- Identify the components of screens and reports to be prototyped.
- Create Prototype Pathways.
- Implement Prototype Pathways.
- Prepare for the prototype demonstration.
- Demonstrate and review the Prototype Pathways.
- Update all appropriate SSADM documentation.

These activities will be explained with reference to a number of examples from the Opera Booking System (see Appendix F).

Define the Prototype Scope

Specification prototyping is based on the demonstration of functions to users. The prototypes consist of screens, reports, and Menus and Command structures.

In general, only critical functions of the system should be prototyped. In this context, a 'critical' function is one that is either of high volume or is of most value to the users of the system. The high-volume functions influence capacity requirements and affect performance, and the 'valuable' functions need to have most attention in design.

In deciding which dialogues and reports to prototype, the following criteria are used:

- The list of critical dialogues documented as part of dialogue identification will form the basis of the list of dialogues to be prototyped.
- Reports are included in the prototyping exercise where:
  — legal constraints are to be imposed on reporting standards;
  — reports are to be used as direct input to another system;
  — reports are of high profile (e.g. used by senior management or members of the public).

Non-critical areas can be prototyped if this is felt to be useful. For example, if the developers are unfamiliar with the prototyping tools, or the users are unfamiliar with computer systems, it may be useful to start the prototyping activity with simple functions, progressing to the critical functions only when all participants are ready.

Where possible, the scope of the prototyping exercise should encompass a full subsystem. The Required System Data Flow Model may be useful in identifying discrete areas of the system that might be most appropriate as a choice for prototyping, but this should be checked with the users who may have a different perception of what they would like to see prototyped.

Prototype Menu Structures and Command Structures

In some cases, it may be useful to design Menu and/or Command Structures at this point. A Menu or Command Structure will give the context of individual dialogues and reports and demonstrate to the users how navigation between functions will be achieved. The work done here may also be useful as an input to the development of standards for inclusion in the Application Style Guide during Technical System Options.

A single Menu or Command Structure is developed for each User Role that has access to some or all of the functions in the chosen area to be prototyped. This is done using the User Role/Function Matrix for reference. It may be valuable to include in the Menu or Command Structure all functions in the chosen area to which each user role will have access, not just the ones that are to be prototyped. This will help to check that the developers and users are in accord on the set of functions allocated to each user role on the User Role/Function Matrix.

Menu Structures can be developed directly using the prototyping tool. The formal documentation of Menu and Command Structures is done during dialogue design in the Logical System Specification Module. The Menus and Command Structures developed during prototyping are more informal but will be used as an input to their formal definition in the subsequent module.

Any amendments required by users should be checked against the complete User Role/Function matrix to ensure consistency across the system. Amendments to the Requirements Catalogue may also be required as a result of this phase of the prototyping.

Identify the components of screens and reports to be prototyped

The prototyped Menu and Command Structures define the context for each dialogue and report to be prototyped from a 'top-down' viewpoint. This activity now works 'bottom-up' to define the building blocks for each dialogue and report.

The basic components of each dialogue and report are groups of data items. The data items for each function are found on the I/O Structure for that function. The relevant I/O Structures are used as an input to defining the prototyping components in the following way:

- The I/O Structures to be used as an input to dialogue prototyping are examined to identify logical groupings of dialogue elements (LGDEs) in anticipation of the dialogue design technique in Logical Design.
- The I/O Structures to be used as an input to prototyping reports are examined to extract only the output portions of the structures.

Create Prototype Pathways

A Prototype Pathway combines elements of the 'top-down' and 'bottom-up' views for a user role into a coherent prototype design for a single 'session'. A single Prototype Pathway is developed for each function that is to be prototyped. The design indicates the sequence of the navigation path to be followed through the prototyping session from the selection of a menu option through to the completion of the function. The pathway should not be a complete dialogue design and should concentrate solely on what is to be demonstrated to the users.

A Prototype Pathway diagram is generated for each proposed Prototype Pathway. The diagram uses the following notation:

- boxes represent a component of the pathway;
- lines indicate the sequence of the navigation between the boxes.

The pathway components might be:

- menus;
- dialogues;
- Logical Groupings of Dialogue Elements;
- reports.

These components are combined into a single diagram that is used as the basis for the development of the prototype.

An example of a Prototype Pathway diagram is shown in Fig. 10.1. Here, the pathway is initiated at the first box at the top of the diagram and proceeds through the boxes listed below in sequence. It should be noted that this does *not* represent a full design of a dialogue. It is a very much simplified version of a dialogue and should *not* contain the following:

- All possible paths through the dialogue. The pathway should concentrate only on the paths that are to be demonstrated to the user and will probably represent a small proportion of the options that will be available in the final dialogue.
- Validation and error handling (unless of prime importance to the application). The pathway should concentrate on the main purpose of each function and not dwell on the 'ifs' and 'maybes'. The purpose is to verify requirements not to enter detailed design discussions.
- All possible messages and warnings that could be generated by the system. The pathway should concentrate on the main features of the dialogues and not cater for all possible system responses.

Implement the Prototype Pathways

This activity involves developing the individual Prototype Pathways into working models to demonstrate to the user on the chosen prototyping tool. Whatever the tool to be used, there are several principles that should be applied to the development of the prototyping pathways:

- The screens and reports should mirror current installation standards (possibly documented in an Installation Style Guide). This will give the prototype the 'feel' of the final system and will not contradict standards that the users may be familiar with from other systems.
- All data item names and formats should conform to definitions in the Data Catalogue. This will help to ensure that the data catalogue definitions are correct.
- Screen and report layouts should be easy to follow.
- Screen and report content should map directly onto the I/O Structures used as their basis. Screens may contain one or more whole Logical Groupings of Dialogue Elements.

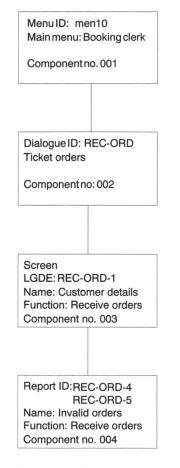

Menu ID: men10
Main menu: Booking clerk

Component no. 001

Dialogue ID: REC-ORD
Ticket orders

Component no: 002

Screen
LGDE: REC-ORD-1
Name: Customer details
Function: Receive orders
Component no. 003

Report ID: REC-ORD-4
        REC-ORD-5
Name: Invalid orders
Function: Receive orders
Component no. 004

Figure 10.1   Example of a Prototype Pathway diagram

Prepare for the prototype demonstration

The demonstration of a Prototype Pathway should be carefully planned. It is difficult to focus attention on the requirements and very easy to get caught up in the minutiae of screen and report design.

A Prototype Demonstration Objective Document is produced for each of the Prototype Pathways. Here, the assumptions made and questions to be asked are listed against each of the pathway components. It should be clear from the outset that the purpose of the prototype is to check the components of the Requirements Specification developed to date and to elicit requirements for presentation and standards. It should not concentrate on precise details of design.

If possible, demonstration data should be generated based on valid data ranges and formats from the Data Catalogue. This will make the system more realistic. If an underlying database has been generated, this can be used to check that data manipulations correspond to the requirements as results can be tested against input values.

The Prototype Pathways should be reviewed together to ensure that one pathway does not invalidate the data for another.

Demonstrate and review the pathways

Each Prototype Pathway is demonstrated to the appropriate User Role(s). It should be made clear that this is a model and is not designed to represent the presentation of the system as it will be in the final implementation. The demonstration addresses the assumptions and queries listed in the Prototype Demonstration Objective Document.

A Prototype Result Log is produced for each pathway demonstrated; this documents the outcome of the demonstration together with any follow-up, in the form of amendments to the existing SSADM documentation required. The changes requested should be prioritized to help in deciding whether to incorporate the changes into the prototype and restage the demonstration or whether simply to incorporate the changes into the SSADM documentation without further work being invested in the prototyping exercise.

Update all appropriate SSADM documentation

All SSADM documentation products that need updating as a result of the prototypes should be amended once the results of the prototyping exercise have been decided and documented. The products that can be updated as a result of the prototyping are:

- Requirements Catalogue;
- User Role/Function Matrix;
- Required System Logical Data Model;
- Data Catalogue;
- Entity Life Histories;
- Effect Correspondence Diagrams;
- Enquiry Access Paths;
- Function Definitions;
- I/O Structures.

Also, it is possible that Menu and Command Structures have been produced. These are not a formal product of specification prototyping but they should be retained for input into the dialogue design activity of Logical Design. Similarly, the I/O Structures annotated with Logical Groupings of Dialogue Elements will form a useful input to dialogue design.

10.6 Considerations for prototyping

Prototyping is optional within SSADM. It will not be appropriate for every project. The costs are significant, both in terms of time and computing resources, and it must be determined from the outset whether the expected results can justify the expenditure. If a decision is made to embark on a prototyping exercise, the choice of tools to develop and demonstrate the prototypes requires careful consideration.

Prototyping suitability

Whether or not prototyping is applicable is determined by a project's characteristics. Prime among these is the value of a project. If a system is of high value, either in monetary or business terms, there may be justification for investing in prototyping.

Other factors that contribute towards the decision to invest in a prototyping solution include the following:

- *User experience*. The users may not be familiar with computer systems and have difficulty in expressing their requirements in an appropriate way and comprehending the implications of the SSADM documentation. Conversely, users may be very bound up in an existing computer system and may not be able to envisage alternative ways of solving their needs. In either case, a prototype can help to animate the ideas of the developers to give users a clear understanding of what is being proposed. Also, if the system will require major changes in working practices, these may need to be tested out before proceeding to full development.
- *Analyst understanding*. The analysts may have difficulties in understanding the requirements for the new system because of their inexperience of the business, a wide user base, which gives conflicting input, or the unavailability of expert users. Here, a prototype can soon demonstrate whether or not the requirements have been understood sufficiently well and it can be used to resolve conflicts. In this case, a prototype may need to undergo a number of iterations before agreement is obtained.
- *High risk/criticality*. If there are high risks, e.g. large sums of money or business success, associated with the project, it is very important to ensure that the requirements have been understood and that users are confident that the right system will be developed. Also, the user interface may be critical to the success of the system, e.g. where members of the public are to use the system or where throughput is high, so specific areas may be prototyped to optimize these aspects.
- *Variety of design solutions*. If there are many possible design solutions, the use of prototyping may help to narrow the options by demonstrating a number of alternatives and seeking preferences from the users and exploring best directions for the developers. This may be particularly relevant where there is no current system to act as a reference.

Project characteristics that might be considered to make a project unsuitable for the use of prototyping include:

- *Well understood requirements*. If the requirements are well understood, e.g. where the new system is a direct copy of another system, then prototyping will not be expected to give sufficient benefit to offset the costs.
- *Costs/timescales*. If the costs must be kept to a minimum or if the timescales are tight, it may be necessary to preclude prototyping and accept that the requirements may not be as accurately modelled as they could be.

Prototyping tool considerations

The choice of software and hardware to develop the prototypes is often determined by what is available. If the target implementation environment is already in place, there are advantages in using it to develop the prototypes to the same standards and using the same tools that will be used during implementation. This may not be possible for several reasons:

- the 'live' computer does not have the capacity to support development work;
- there may not be facilities readily available where the developers are located;
- the final solution may use technology that is difficult to use and be inflexible in implementing changes.

In 'green-field' situations, the implementation hardware and software will not yet have been selected at the point where prototyping becomes applicable. Therefore, the majority of projects planning to undertake prototyping will need to determine the best tools for the job. There are two conflicting factors — *cost* and *sophistication* — in determining which tools to select, and these two factors need to be carefully balanced, taking into consideration the importance of the project and its associated risks.

The degree of sophistication of the tool influences the degree to which the prototype can demonstrate the facilities of the final system, for example:

- If a simple tool is used, it may only be possible to paint screens that are static.
- If an application generator is used, a simple underlying database can be developed which can be used to store and retrieve records to animate the prototype.

There are advantages and disadvantages associated with these two extremes:

- A simple system has the advantage of being quick to develop and easy to use. It also has the advantage of not being convincing as a 'real' system so there is less danger of raising false expectations about development timescales!
- The disadvantage of a simple system is that it will be unrealistic and may communicate the wrong impression to the users. Also, there is more of a burden placed on the developers in demonstrating the prototype to make it appear like a real system.
- The advantage of a more sophisticated system is that the users will get the feel for a working system and may be able to contribute more to the discussion on what is required.
- The disadvantage of the sophisticated system, apart from high cost, is that it may take longer to develop and may raise user expectations about its readiness for use.

The features of a support tool that should be considered are:

- screen, menu, and report painting;
- on-line navigation;
- database;
- active data dictionary;
- application logic simulation;
- application data storage simulation;
- version control.

The support tool chosen should allow easy development and changes to prototypes and it is useful if its facilities match the facilities expected of the target system so that the user can see what the final system will look like. Alternatively, some computer-aided software engineering tools are beginning to incorporate simple prototyping tools which may be of use.

10.7 Comparison with Version 3

SSADM Version 3 does not formally incorporate prototyping into the steps and stages, though some general advice is included on the possible use of prototyping to support SSADM. In Version 4, a more formalized approach has been taken with the inclusion of specification prototyping as a step within the Requirements Specification Module.

11. Effect Correspondence Diagrams and Enquiry Access Paths

11.1 Introduction

Effect Correspondence Diagrams and Enquiry Access Paths are not generally grouped together within SSADM Version 4 structure and documentation:

- Effect Correspondence Diagrams are covered by entity-event modelling.
- Enquiry Access Paths are covered by logical data modelling.

They have been grouped together here because they share a similar purpose within the method in that they both form the basis for the process models developed during logical database process design in the Logical Design stage. The Effect Correspondence Diagrams and Enquiry Access Paths can be thought of as defining the 'what', while the process models define the 'how'. In addition, both techniques can be used to validate further that the Required System Logical Data Model satisfies the processing requirements of the system. Both techniques can be regarded as stepping stones between the functional aspects of the Requirements Specification Module and the logical database process design in the Logical System Specification Module.

The two techniques have a similar notation in that they both use some elements of the standard structuring technique described in Appendix D. Each technique uses the concept of selection and iteration and represents accessed entities. The major difference between them is that Enquiry Access Paths place more emphasis on the sequence of entities accessed and reflect the navigation path through the Required System Logical Data Model (using arrows to show direction of navigation), whereas Effect Correspondence Diagrams represent a more static view of relationships between entities accesses (using double-headed arrows to show correspondences).

As there are slight differences in derivation and notation, the two techniques are treated separately within this chapter.

11.2 Naming conventions

An *Enquiry Access Path* reflects the navigation through the Required System Logical Data Model needed to support an enquiry function of the system or an enquiry fragment of an update function.

The diagram is built up using a series of *access correspondences*, *optionality*, and *iteration* and can be based on a *required view* of the Required System Logical Data Model.

An *Effect Correspondence Diagram* is produced for each of the events within the system.

The diagram is built up using a series of *one-to-one correspondences*, *optionality*, and *iteration*.

11.3   Place in Structure

Both Enquiry Access Paths and Effect Correspondence Diagrams are first developed in Step 360: Develop processing specification. As a result, both products are part of the Requirements Specification.

Based on practical experience, it may sometimes be considered preferable to produce Enquiry Access Paths immediately prior to the production of Enquiry Process Models (Step 530) and Effect Correspondence Diagrams immediately prior to Update Process Models (Step 520). In this case the Required System Logical Data Model is not fully validated until Logical Design.

11.4   Place in Product Breakdown Structure

Enquiry Access Paths are one of the constituent parts of the Function Definitions. Function Definitions appear twice within the Product Breakdown Structure:

- within the Processing Specification in the Requirements Specification;
- within the Logical Process Model, which is part of the Logical System Specification.

Effect Correspondence Diagrams also appear in both the Requirements Specification and the Logical System Specification. This is shown in Fig. 11.1.

11.5   Enquiry Access Paths

An Enquiry Access Path is produced for each enquiry function. It defines the entities accessed and the navigation path through the Required System Logical Data Model that are needed to support the enquiry function. If required, Enquiry Access Paths may also be used to model the enquiry components of an update function.

Enquiry Access Paths are built up from the Required System Logical Data Model during the Requirements Specification module. The notation for Enquiry Access Paths contains some of the elements of the standard structure diagrams (see Appendix D) such as selection and iteration. However, they do not conform to the full syntax defined there and should *not* be drawn with reference to the standard structuring rules.

Enquiry Access Paths are used to:

- validate that the Required System Logical Data Model can support the enquiries;
- act as a stepping-stone in the definition of the Enquiry Process Models;
- document the detailed requirements for the enquiry in terms of the entities accessed.

Enquiry Access Paths are built up with close reference to the Required System Logical Data Model and are developed in parallel with entity–event modelling within Step 360:

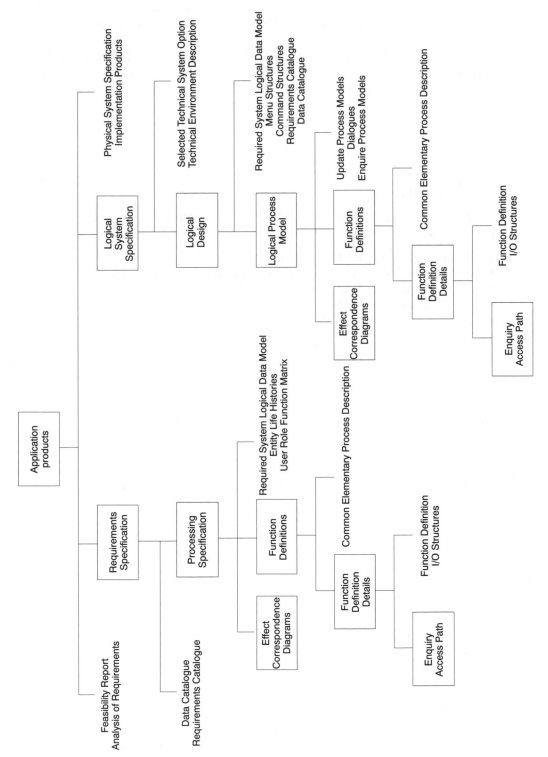

Figure 11.1 Partial Product Breakdown Structure

123

Develop processing specification. Alternatively based on practical experience, it may be considered preferable to produce the Enquiry Access Paths immediately before producing the Enquiry Process Models.

Before an Enquiry Access Path can be developed, the trigger for the enquiry should be determined. The enquiry trigger will consist of any information that must be supplied to allow the enquiry to select the correct occurrences of data for output. The enquiry trigger data will be defined by (or be a subset of) the data on the input leg(s) of the I/O Structure for the enquiry function.

The data that triggers an enquiry will normally be one, or a combination, of:

- prime key (or keys) of any entity to be accessed directly by key;
- non-key attributes which will produce multiple accesses of one or more entities;
- selection criteria to be used.

After naming each enquiry and establishing the enquiry trigger, each Enquiry Access Path is built up in a number of tasks:

- Entities containing the required data are identified.
- A 'required view' of the Required System Logical Data Model is drawn.
- The required view is converted into an Enquiry Access Path.
- The entry point(s) for the enquiry are identified and the input data items are added.
- The path is checked to ensure all required data may be obtained.
- The Logical Data Model is annotated to show entry points.

These tasks will be explained with reference to a number of examples from the Opera Booking System (see Appendix F). The Required System Logical Data Structure for the Opera Booking System is shown in Fig. 11.2 for reference.

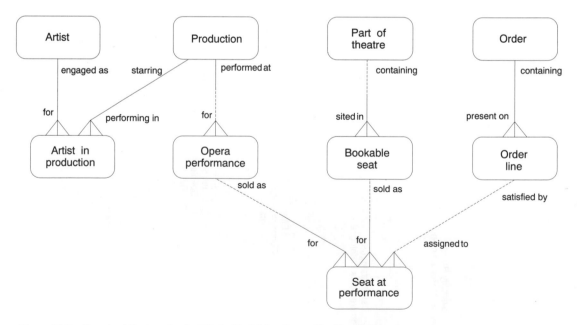

Figure 11.2   Required System Logical Data Model for Opera Booking System

Identify required entities

All data items identified on the I/O Structure as being output from the enquiry must:

- *either* come from entities on the Required System Logical Data Model;
- *or* be derived by processing.

The data content of the entities should be examined to find which entities must be accessed for this enquiry.

(Note that if Relational Data Analysis has been done using the I/O Structures from the functions, it is unlikely that any data items will be missing at this point as these will have been incorporated as part of Step 340.)

In the Opera Booking System, one of the enquiries lists all artists involved in a specific day's performance who are being paid an hourly rate in excess of £50. This enquiry must access the 'opera performance' entity for the date of the performance, the 'artist in production' entity to find the hourly rate and the entity 'artist' to obtain the performers' names. There may be a need to access other entities for navigational purposes, but these will be identified in the next few tasks.

Draw 'required view' of Logical Data Model

Using the entities identified above, the Required System Logical Data Model should now be examined to find out how the enquiry is going to navigate around the data structure.

If it is felt to be useful, the relevant part of the Logical Data Structure may be redrawn to show the way it will be used in the enquiry. This partial structure is called a 'required view' and is drawn so that any accesses from master to detail are drawn vertically and accesses from detail to master are drawn horizontally. This will require an identification of entities that are required to navigate between the entities containing the data to be output so that the required occurrences of each entity are selected.

The required view of the Logical Data Structure needed to support the enquiry listing all artists performing on a specific date is shown in Fig. 11.3. This diagram shows that the entity used to initiate the enquiry is 'opera performance'. Each opera performance that takes place on the specified date will need to navigate through the 'production' entity and 'artist in production' entity to find all the relevant artists who will be performing on that date, within the hourly rate range specified.

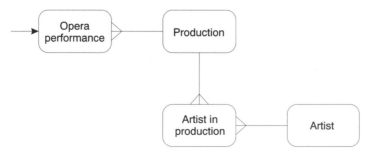

Figure 11.3   Required view of Logical Data Structure for artists performing on a specified date

If while drawing up the required view it becomes apparent that the enquiry is not supported by the Logical Data Model, then either the Logical Data Model should be altered or a processing solution (e.g. a sort) should be considered.

Convert required view into an Enquiry Access Path

The required view is enhanced to become the Enquiry Access Path by adding in some indication of the way in which the enquiry will be processed. This is done by explicitly showing the sequence of access and indicating where there will be selections and iterations. The diagrammatic technique includes: *access correspondence arrows*, *iteration*, and *selection*.

ADDING ACCESS CORRESPONDENCES

An access correspondence arrow is used to indicate that the access of a single occurrence of one entity corresponds to an access of a single occurrence of connected entity. For example, if the access is from detail to master, there will only be one occurrence of the master entity for each detail accessed. The arrow shows the direction of access.

Similarly, an access correspondence can be annotated on a master to detail accesses where only one occurrence of the detail is required for each occurrence of the master. This could happen when an entity contains historical information but only the latest occurrence is required.

ADDING ITERATIONS

An iteration is added where more than one occurrence of an entity is required for a single occurrence of a connected entity. For example, an access from master to detail will almost always involve an iterated access.

An iteration is denoted by the addition of:

- an extra box, which represents the complete set of repeated accesses;
- an asterisk to the entity that is accessed repeatedly;
- a correspondence arrow between the entity accessed once and the 'set of' box above the repeating entity.

This is demonstrated by an example of an Enquiry Access Path for a function to list bookable seats and seats at performances for a specified part of the theatre in Fig. 11.4. Here, a simple enquiry is needed to list all seats that may be booked and all seats at performances which are in a particular area of the theatre. This enquiry simply navigates down two master-to-detail relationships ('part of theatre' to 'bookable seat' and 'bookable seat' to 'seat at performance'). There are two iterated accesses required which necessitates the addition of two 'set of' boxes as shown.

ADDING OPTIONALITY

Optionality is added to the diagram where a split in the access path occurs and a choice is made between one or more subsequent paths based upon specified criteria.

As with standard structuring techniques, optionality is denoted by a circle in the top of each of the possible selections. The selections can either be accesses in their own right or, more commonly, a statement of the criteria applied to the selection.

The Enquiry Access Path for the enquiry to list all artists for a specific performance is shown in Fig. 11.5 and illustrates the addition of both iteration and optionality. Here,

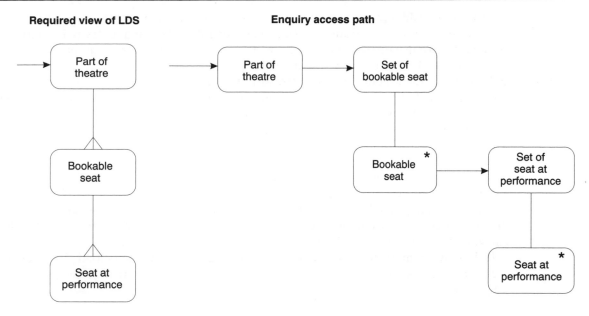

Figure 11.4 Enquiry Access Path for all seats for a specified part of theatre

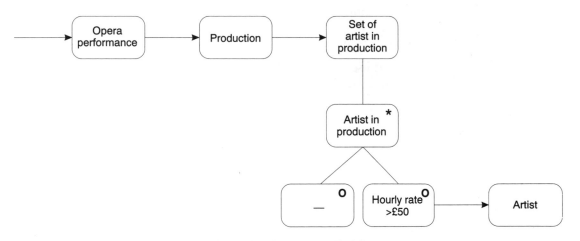

Figure 11.5 Enquiry Access Path for artists performing on a specified date

access correspondences are used for the detail-to-master accesses and an iteration is used for master-to-detail access a selection is added to show that only artists that are on an hourly rate of more than £50 are accessed. It is worth noting that if the entry point had been on 'artist' instead of 'opera performance', 'opera performance' would be iterated as well as 'artist in performance' and the diagram would be reversed.

Identify entry points and add input data items

The construction of each Enquiry Access Path identifies the initial entry point into the enquiry, shown by an arrow pointing to the first entity access. These entry points should now be defined further and any additional entry points identified.

Each entry point should be annotated with the essential data required for that access. For example, if the prime key of an entity is used for access, then the data item(s) constituting that key should be added to the line. Figure 11.6 shows the example in Fig. 11.4 updated to show the prime key of the entity 'part of theatre' being used for access.

It is also possible that a key other than the prime key of the first entity will be used for access. Figure 11.7 shows the Enquiry Access Path from Fig. 11.5 updated to show that 'date of performance' is the item used for access.

As the data item used to access the entity is not its prime key, it is possible that more than one occurrence will be selected. This can be re-expressed using a 'set of ' box at the entry point as shown in Fig. 11.8.

If there is a requirement to read all occurrences of an entity, then either the arrow to the entry point is left unlabelled; or alternatively an unlabelled arrow is shown for the 'set of' box.

Check the path

The Enquiry Access Paths should be checked to ensure that all required data can be obtained from the entities using:

- direct reads to the prime keys;
- read next detail of a detail from the master;
- read directly master from detail.

In addition it would be useful to check that any data which needs to calculated has all the necessary items needed for the calculation.

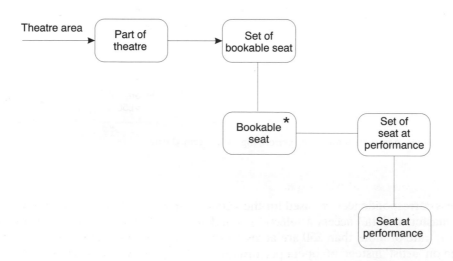

Figure 11.6 Enquiry Access Path from Fig. 11.4 updated to show access key

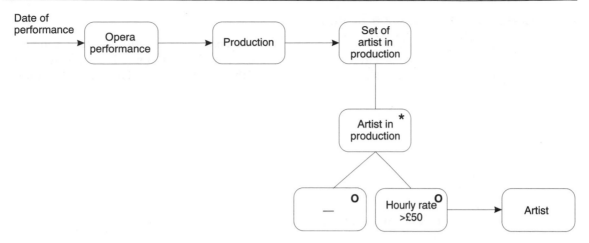

Figure 11.7 Enquiry Access Path from Fig. 11.5 updated to show access key

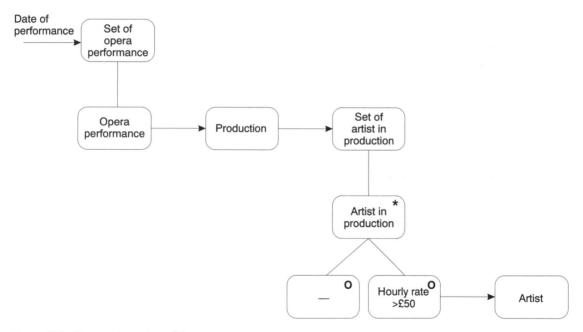

Figure 11.8 Entry point as 'set of' box

Show entry points on the Logical Data Model

It may be a useful input to Physical Design if all the entry points identified on Enquiry Access Paths are annotated on the Required System Logical Data Model. If this is done, the entry points should be indicated by arrows with the data used for entry on the Required System Logical Data Structure Diagram. This will give an indication of which entities will need to be accessed directly in the physical system.

11.6   Effect Correspondence Diagrams

Effect Correspondence Diagramming is the second part of entity–event modelling, following on from — and validating — Entity Life History analysis. An Effect Correspondence Diagram is produced for each of the events identified and modelled on the Entity Life Histories.

Each Effect Correspondence Diagram models the entity accesses required to support an event. While an Entity Life History shows all the events that affect an entity, Effect Correspondence Diagrams show all the effects on entities from a single event. They document diagrammatically all the effects of a particular event on entities and include read accesses to entities for validation or navigation.

Effect Correspondence Diagrams are similar to Enquiry Access Paths in that they model one-to-one correspondences, selection, and iteration. They are dissimilar in that they do not always reflect the sequence of access and need not strictly show the complete navigation path.

Effect Correspondence Diagrams are used for the following:

- To validate the Entity Life Histories. This is particularly useful as analysts and users sometimes find it easier to focus on an event rather than trying to visualize the entire life of an entity.
- To act as a stepping stone in the definition of Update Process Models developed during Logical Design.
- To define the detailed requirements for the event in terms of the entities updated by the event.

Each Effect Correspondence Diagram can be constructed using a number of tasks:

- a box is drawn representing each effect of the event;
- simultaneous effects are identified and annotated on the diagram;
- optional effects are identified and annotated;
- iterative effects are identified and annotated;
- one-to-one correspondences are added to the diagram;
- related iterative effects may be merged;
- entities that are not updated are added;
- event data is added.

Each of the above steps produces an intermediate diagram building up to the final Effect Correspondence Diagram. It should be noted that it is the final diagram that is required and the others should be regarded as 'working models' used to derive the Effect Correspondence Diagram. Also, not all the steps are necessary for all Effect Correspondence Diagrams. These activities will be illustrated with reference to the Opera Booking System (see Appendix F).

Draw a box for each entity affected

As a starting point for an Effect Correspondence Diagram, it is necessary to identify all the effects of the event that may appear on a number of Entity Life Histories. It has been found useful in practice to produce an intermediate document at this stage which catalogues each event with all its effects together with the effect qualifiers and/or entity roles. This is a useful reference document to use while building up the Effect Correspondence Diagrams.

For the Opera Booking System, the event 'arrival of order' appears on three Entity Life Histories, namely 'order', 'order line', and 'seat at performance'. These three entity names are placed in boxes to start the diagram as shown in Fig. 11.9.

Identify simultaneous effects

Simultaneous effects occur where a single occurrence of an event affects more than one occurrence of a particular entity in different ways at the same time. In this case, the Entity Life Histories show the effects qualified by an entity role name to indicate the different ways in which the event acts. A common example of simultaneous effects is where one occurrence of an entity is replacing another. The event that causes this replacement creates one occurrence of the entity while deleting another occurrence.

Each simultaneous effect is drawn as a separate 'soft box'. The entity name is put in each of the boxes qualified by the entity role name taken from the Entity Life History. The fact that these are simultaneous effects and therefore happen together is denoted by enclosing them in a larger 'soft' box. This can be seen with reference to another Effect Correspondence Diagram from the Opera Booking System for 'performance details finalized', which is shown in Fig. 11.10. Here, an artist can be replaced by another artist if the production details are changed. A single occurrence of the event creates a new occurrence of the entity 'artist' at the same time as deleting the replaced occurrence of 'artist'.

Identify optional effects

Optional effects occur in one of two ways. First, if a single occurrence of an event can affect an entity in two or more mutually exclusive ways, then the effects are classified as optional on the Effect Correspondence Diagram. In this case the Entity Life History shows the effects qualified by effect qualifiers to indicate the different ways in which the event acts. A common example of optional effects is where the same event can affect the entity at different points in its life, for example both before and after authorization. Thus, if an Entity Life History contains entity roles, these will become simultaneous effects on the Effect Correspondence Diagram, and if it contains effect qualifiers they will become optional effects.

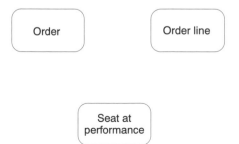

Figure 11.9   Activity 1 of Effect Correspondence Diagram construction

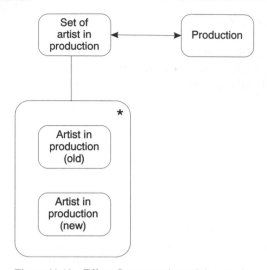

Figure 11.10 Effect Correspondence Diagram for 'performance details finalized'

The other situation where optional effects occur is if the effect can either happen or not within the context of the event. A common example of this might be where an invoice is marked as finished only if all the invoice lines are paid off.

Optional effects are added to the diagram using the standard notation for selection as shown in Fig. 11.11. Here, the entity name is put in each box under the selection and annotated with the effect qualifier from the Entity Life History. The event 'arrival of order' causes the creation of 'order line' in two possible ways on the Entity Life History — once in the context of a satisfied order and once in the context of an unsatisfied order.

Show iterative effects

By now, all the effects of an event should be shown somewhere on the diagram and it is possible to start to think how each effect relates to other effects. Before links between effects can be identified it is necessary to decide whether each effect will occur for a only one occurrence of the entity or for a number of occurrences. If it is for a number of

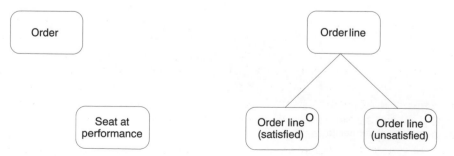

Figure 11.11 Activity 3 of Effect Correspondence Diagram construction

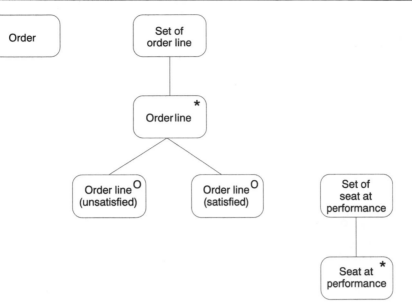

Figure 11.12 Activity 4 of Effect Correspondence Diagram construction

occurrences, there will be an iteration, which is denoted by an asterisk in the top right corner of the effect box and a 'set of' box placed above it as shown in Fig. 11.12. In this example, 'arrival of order' may affect more than one occurrence of the entity 'seat at performance' because an order can contain bookings for more than one seat at performances. An order will also contain more than one order line.

Add one-to-one correspondences

The boxes on the diagram are now connected to indicate where there is a relationship, or correspondence. The only type of correspondence shown is a one-to-one correspondence. This means that if a box on the diagram occurs one-to-one with another box on the diagram, they are connected with a double-headed arrow. This is why the 'set of' boxes have been put on the diagram: a single occurrence of an effect on a master entity usually has a one-to-one correspondence with a set of its details. The addition of one-to-one correspondences is demonstrated in Fig. 11.13.

Correspondences do not strictly have anything to do with the Required System Logical Data Model. They are determined from the event's view of how effects relate to one another. However, if this principle is taken to extremes, it can become very confusing when trying to determine correspondences. For example, if there is one effect that has a single occurrence and ten that are iterative, how are correspondences determined? What is to prevent the analyst from drawing a correspondence from the single occurrence directly to all the 'set of' boxes? If, on the other hand, reference is made to the Logical Data Model, it is probable that a hierarchy of correspondences can be identified where the master–detail relationships will help to determine sensible correspondences. This is illustrated in Fig. 11.14.

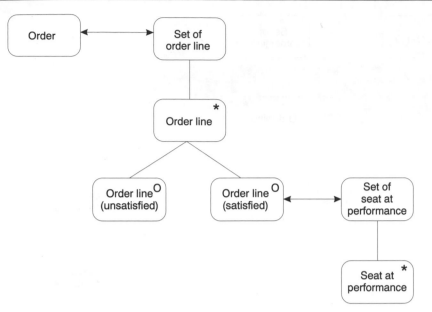

Figure 11.13   Activity 5 of Effect Correspondence Diagrams

In general, experience of Effect Correspondence Diagramming has shown that it is useful to refer to the Required System Logical Data Model to determine correspondences, even to the extent of considering navigation through the model. This makes the technique more 'meaningful' and therefore easier to use. It can also assist the physical design activities in showing how the Logical Data Structure is navigated for the event.

Merge associated iterative effects

The diagram is checked to see if there are any interrelationships between iterated effects that have not been identified so far. This is unlikely to happen if the activities are followed strictly. However, if iterations have been identified before selections, there may be a situation where the same detail entity name appears in effects below several 'set of' boxes which correspond to the same effect on a master entity. In this case, a selection may need to be introduced. This is demonstrated in Fig. 11.15 with reference to how the Effect Correspondence Diagram for the event 'arrival of order' could have been drawn if the selection had not been identified earlier. In the example the selection was not identified and as a result two sets of 'Order Line' effects documented. These should be merged together.

This should be undertaken with extreme care — if there are two iterations of the same entity, this may indicate a correspondence with different occurrences of the master entity. In this case they should be kept separate and the master 'effect' split in two as shown in Fig. 11.16. This is where the Effect Correspondence Diagram technique may lead to a re-examination of the Entity Life Histories and events.

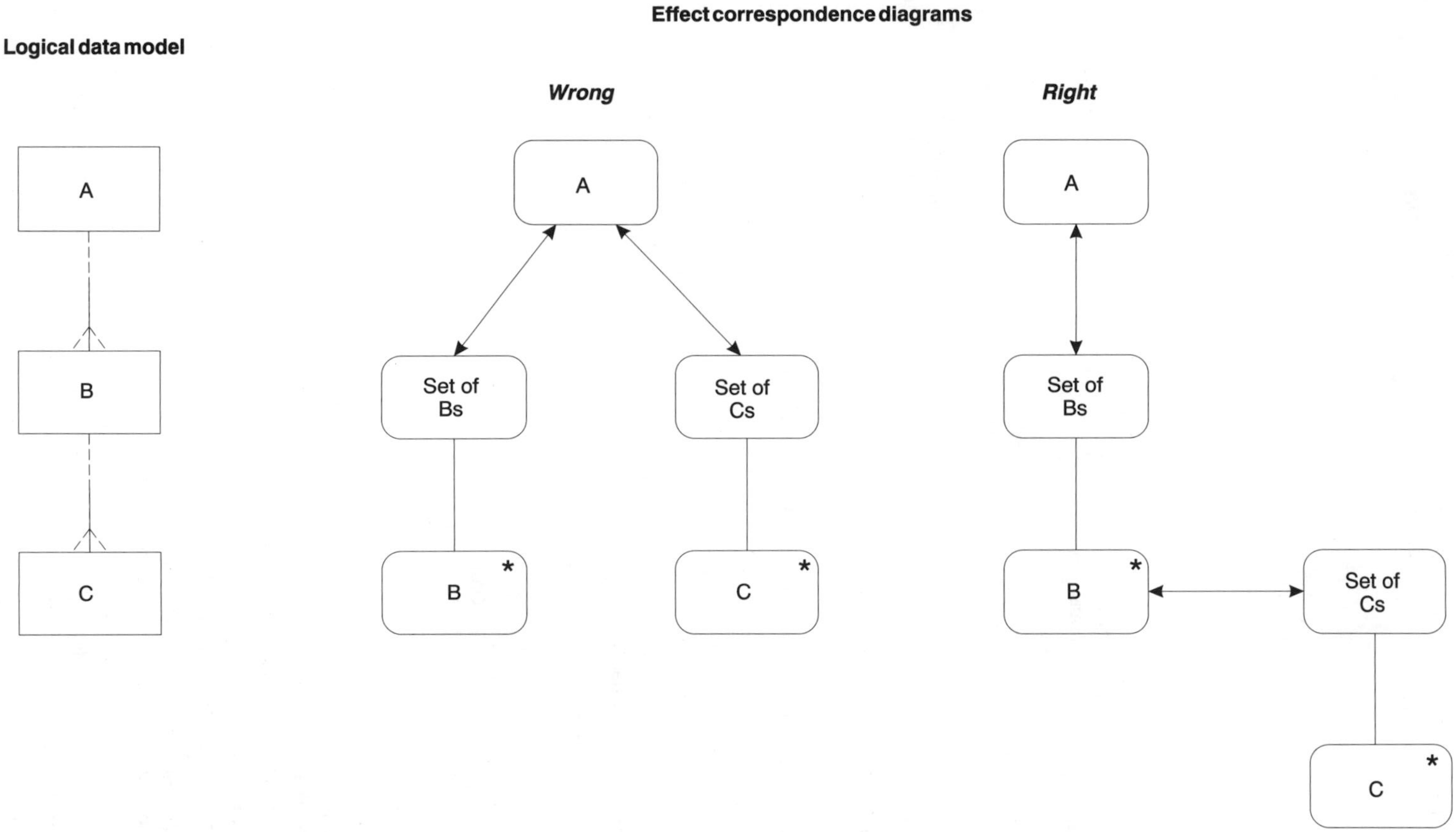

Figure 11.14   Demonstration of how the data model relates to Effect Correspondence Diagramming

135

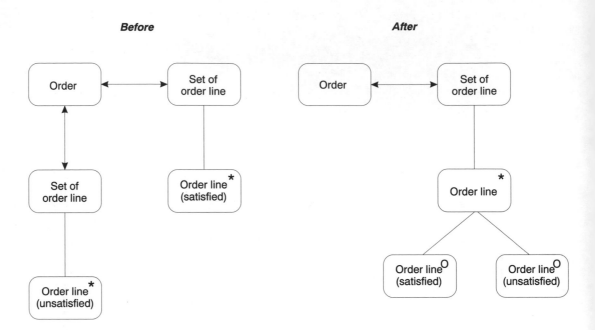

Figure 11.15   Activity 6 of Effect Correspondence Diagrams

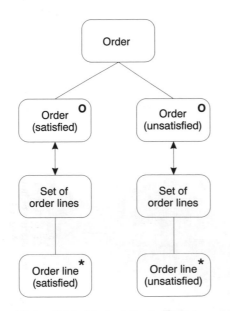

Figure 11.16   Master 'effect' split in two where there are two iterations of the same entity

Add non-updated entities

An entity may need to be accessed by an event without being updated where:

- it is required for navigation; or
- where data is required on a read-only basis either for calculation or the determination of a selection and/or iteration.

If an entity is required for either purpose, it should be added to the diagram as a non-updating 'effect'.

Add event data

The Effect Correspondence Diagram should be annotated with any data that is necessary for the event to proceed. An example of event data is the key of the first entity accessed. This is shown on the diagram with an arrow pointing to the box on the diagram where the event data is required. In many cases, this will show the entry point for the event as demonstrated in Fig. 11.17. Here, the diagram is showing that 'order number' is input by the user to create the order and associated entities. Also the 'seat at performance' identifiers are supplied to enable the system allocate specific occurrences to 'order lines'. It is possible to show all the data items required to create the other entities here, but this would not be particularly helpful and would clutter the diagram. It is most important to show 'key' items that are used for direct access.

'In practice' hints

Like many of the techniques of SSADM, each practitioner can devise a way of using the technique that they can understand and that will work best in their situation. This is a powerful technique that can be of vital importance in checking the completeness of the Entity Life Histories and act as a non-procedural statement of the data usage of an event.

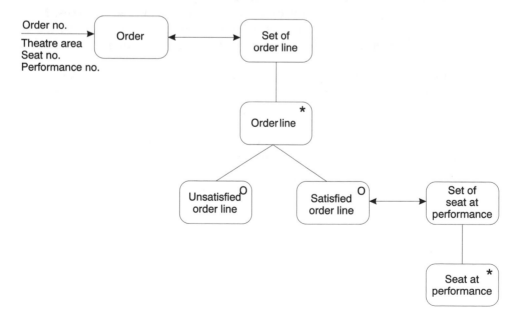

Figure 11.17 Effect Correspondence Diagram with event data added

The activities listed above can be varied if this is considered useful. One way of approaching the construction of an Effect Correspondence Diagram is to 'walk through' the event, working out which entities will need to be accessed and in what order. This can then be used as a context within which to consider individual effects.

In some cases, there may be some validation that needs to be done before the event can proceed. In this case, an Enquiry Access Path can be used, or a fragment of an Effect Correspondence Diagram to show this pre-event validation.

Some other 'tips' that have come out of the use of the technique on projects are:

- All the boxes do not need to join up — there can be different 'phases' of an event, each of which requires event data to be input.
- There may be more than one box requiring event data on the same structure.
- As some boxes are effects (Insert, Modify, Delete), some are reads and some are simply nodes, it is useful to annotate the boxes on the diagram to distinguish which is which.
- It is useful to distinguish between 'direct' and 'indirect' correspondences. A direct correspondence is supported by a relationship on the Logical Data Model and access can be assumed to be via navigation of the relationship. Indirect correspondence is not supported via a relationship and additional event data will be required for the access.
- As navigational entities are included, it is possible to build up the Effect Correspondence in much the same way as an Enquiry Access Path. This should be done with care, however, as the full access path is not defined by an Effect Correspondence Diagram: it is a statement of the 'what' rather than the 'how'.

11.7 Comparison with Version 3

There is no direct equivalent to Enquiry Access Paths and Effect Correspondence Diagrams in Version 3 with the exception of an optional Logical Access Map, which is a subset of the Process Outlines. The Logical Access Map identifies the sequence of accesses through the Logical Data Structure and adds volumetric information to the accesses.

Otherwise, the only equivalent in Version 3 is the access information contained on the Logical Enquiry Process Outlines and Logical Update Process Outlines.

It should be noted that the Process Outlines contain more information than the Enquiry Access Paths and Effect Correspondence Diagrams to do with volumes of entities accessed. If volumetric information is required, then the Enquiry Access Paths and Effect Correspondence Diagrams could be augmented to show the number of each entity that will be accessed.

12. Logical Database Process Design

12.1 Introduction

Logical Design is made up of two parallel streams:

- Dialogue Design.
- Logical Database Process Design.

Each stream is based on elements of the Requirements Specification and is carried out independently of the other stream. This is represented in Fig. 12.1. Dialogue Design concentrates on modelling the user's view of the processing aspects of the system. Logical Database Process Design, which covers the logical design of both enquiries and updates, concentrates on modelling the processing from the system's point of view by looking at the updates and enquiries to the data.

Both updates and enquiries are modelled:

- An Update Process Model is produced for each event.
- An Enquiry Process Model is produced for each enquiry function.

In addition, common processing can be defined using the same modelling technique and referenced by the process models.

The logical process models produced by Logical Database Process Design are used:

- as a basis for Physical Design;
- to provide a definition of the system's processing to be used for future maintenance and enhancement of the system.

Update Process Models and Enquiry Process Models both use the standard structuring conventions which are applied to a number of other SSADM techniques (see Appendix D).

12.2 Naming conventions

Logical Database Process Design (LDPD) models the system processing's interaction with the system's data during Logical Design.

Enquiry Process Models are produced for each of the enquiry functions and are based on *Enquiry Access Paths* and the output leg of the *I/O Structure*.

Update Process Models are produced for each event and are based on *Effect Correspondence Diagrams*.

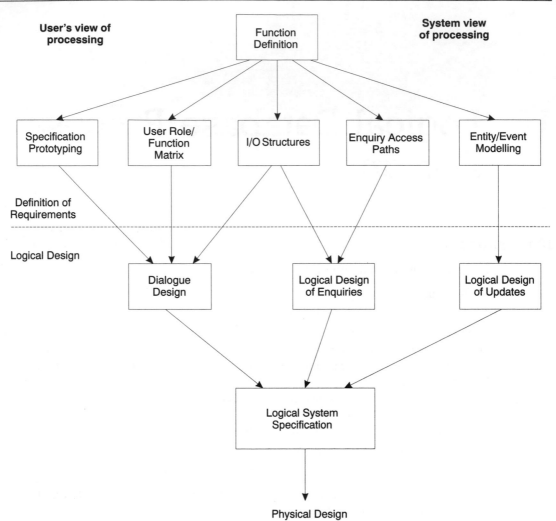

Figure 12.1 Graphic of the Logical Design parallel streams

12.3 Place in structure

Update Process Models are produced in Step 520: Define Update Processes followed by the production of Enquiry Process Models in Step 530: Define Enquiry Processes.

In practice these two steps can be carried out in parallel. However, it should be remembered that either of them could cause an amendment to the Required System Logical Data Model and thus affect the other activity.

12.4 Place in Product Breakdown Structure

Both Update Process Models and Enquiry Process Models are part of the Logical Process Model. The Logical Process Model is, itself, a part of the Logical Design which is a major part of the Logical System Specification. This is represented in Fig. 12.2.

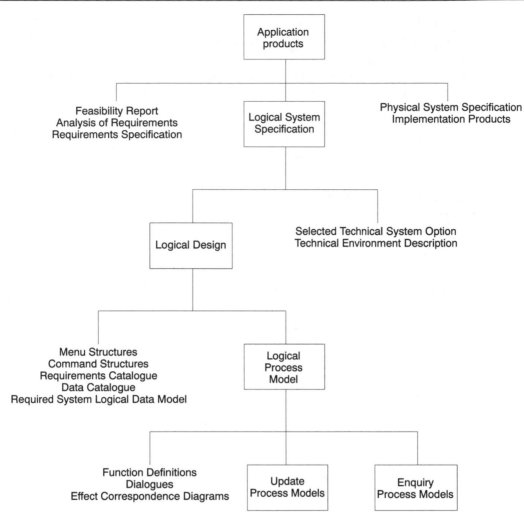

Figure 12.2 Partial Product Breakdown Structure

12.5 Notation and use

Both Update Process Models and Enquiry Process Models are developed during Logical Design in parallel with Dialogue Design. Update Process Models are produced first, based upon Effect Correspondence Diagrams. Following this, Enquiry Process Models are produced, based upon Enquiry Access Paths and I/O Structures. Both types of model use the same structuring conventions, which are based on the standard conventions described in Appendix D.

Update Process Models

An Update Process Model is built up from the Effect Correspondence Diagram for an event by following a number of relatively mechanistic rules. The resulting structure is then enhanced by the addition of operations to give a very detailed specification of the processing required.

Each Update Process Model is built using a number of tasks:

- Effects in one-to-one correspondence are grouped.
- Operations are listed.
- Effect Correspondence Diagrams are converted to 'Jackson-like' notation.
- Extra structure boxes are added to complete the structure.
- Operations are allocated to the structure.
- Conditions are allocated to the structure.
- Integrity error conditions are identified.
- Error outputs are identified.
- The structure is walked through.

These activities will be illustrated with reference to the Opera Booking System (see Appendix F). Here, the Effect Correspondence Diagram for the event 'arrival of order' which was built up in Chapter 11 will be converted into an Update Process Model and is repeated in Fig. 12.3. The event is concerned with creating an order with its associated order lines and linking each order line that can be satisfied to the 'available seat' entity by the creation of a number of 'seat at performance' entity occurrences. (Note that the entities 'opera performance' and 'available seat' do not appear on this diagram, even though relationships to them are being created. This is because it is assumed that relationships are maintained only from the 'detail' end of the relationship, so the masters will not be affected.)

GROUP ONE-TO-ONE CORRESPONDENCES
This Effect Correspondence Diagram is now examined to identify groups of one-to-one correspondences. On this diagram, there are only two one-to-one correspondences

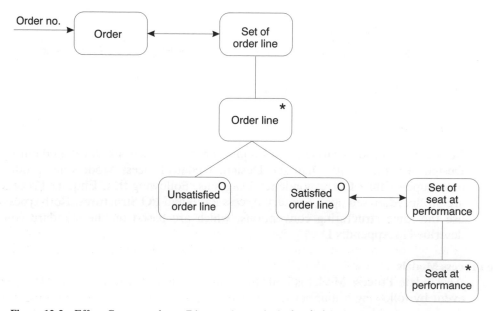

Figure 12.3 Effect Correspondence Diagram for 'arrival of order'

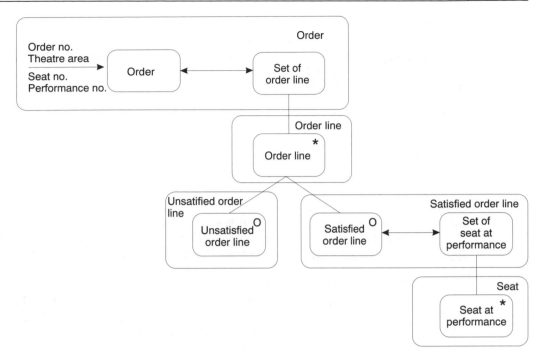

Figure 12.4 Lassoed Effect Correspondence Diagram for 'arrival of order'

between two different sets of effects/nodes. These are between 'order' and 'set of order line' and between 'satisfied order' and 'set of seat at performance'. These are 'looped' or 'lassoed' as shown in Fig. 12.4. Additionally all other boxes on the diagram are also looped individually as a basis for building up the processing structure.

If one-to-one correspondences are between a single box and a large number of other boxes, they are all included within a single 'loop' unless some factor indicates that some should be separate from a number of others. An example of this is where there may be two different 'sets' of one-to-one correspondences and in this case each set may be looped individually.

It is useful if the looped boxes are given a name at this point to identify them. The name should describe the type of processing that will be done for the group.

LIST THE OPERATIONS

Each of the looped boxes represents a piece of processing. Each piece of processing can be further defined in terms of a set of simple operations that will be performed. The initial list of operations can be derived from the appropriate Entity Life Histories. On the Entity Life Histories a number of operations have already been defined to describe the effect. These are used as the basis for the list. The operations from Entity Life Histories are generally only to do with giving values to attributes and maintaining relationships between entities. At this stage, this existing set of operations may need to be enhanced and some additional operation types can be identified:

- reading the entity;
- 'failing' because of invalid state indicator values;
- creating the entity (if this has not already been covered in Entity Life Histories);
- setting the state indicator;
- writing, or storing, the entity onto the database.

In addition, the authors suggest it may be useful to identify any operations to perform calculations. Please note these are not a standard part of SSADM.

The list of operations for the event 'arrival of order' is shown in Fig. 12.5. If this list is compared with the original list added to the 'order' Entity (see Fig. 8.5), you can see that some of the operations descriptions have been enhanced, for example 'store keys' is now 'store keys of *order*'. Another type of enhancement might be to specify values to which attributes may be set. Note the inclusion of the extra operations. An example of this is operation 2, which tests that the order is not already present within the system.

Although not used until later, it is best to identify the operations at this stage as the individual entities affected are visible from the 'looped' diagram illustrated in Fig. 12.4.

CONVERT TO 'JACKSON-LIKE' NOTATION

It is now possible to convert the Effect Correspondence Diagram into a Jackson-like processing structure. The first activity towards achieving this is to apply a number of rules mechanistically to each 'looped' Effect Correspondence Diagram to convert it into an initial structure. This is done as follows:

- The entry point for the event is taken as the starting point.
- The 'loop' containing the entry point is represented by a single box at the top of the diagram with a suitable name. The loop called 'order' in Fig. 12.4 is represented by a box called 'process order' as shown in Fig. 12.6.
- Adjacent loops are now dealt with.

Operations List

1. Read ORDER by key
2. Fail if State Indicator of ORDER <> null
3. Create ORDER
4. Store keys of ORDER
5. Store remaining attributes of ORDER
6. Set State Indicator of ORDER to 1
7. Write ORDER

8. Create ORDER LINE
9. Store keys of ORDER LINE
10. Store remaining attributes of ORDER LINE
11. Store Satisfied Order Line Indicator of ORDER LINE using (Satisfied Order Line = s)
12. Store Satisfied Order Line Indicator of ORDER LINE using (Satisfied Order Line = f)
13. Set State Indicator of ORDER LINE to 1
14. Tie ORDER LINE to ORDER
15. Write ORDER LINE

16. Read SEAT AT PERFORMANCE by key
17. Tie SEAT AT PERFORMANCE to ORDER LINE
18. Write SEAT AT PERFORMANCE

Figure 12.5 Operations list for 'arrival of order'

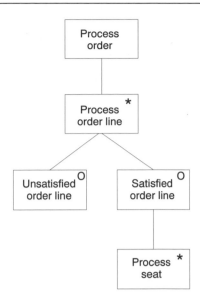

Figure 12.6 Initial process model for 'arrival of order'

- If the loop is adjacent because there is an iteration, an iteration is placed below the first box. In Fig. 12.4, the only adjacent loop is 'order line', so this is shown as a 'process order line' below the 'process order' box in Fig. 12.6.
- If the loop is a selection, a selection is then shown on the process structure. In Fig. 12.4, the loops adjacent to 'order line' are 'unsatisfied order line' and 'satisfied order line'. These are shown as a selection under the iteration as shown in Fig. 12.6.

The diagram is completed by showing the iteration of 'seat' under 'satisfied order line'.

ADD STRUCTURE AND SEQUENCE BOXES

The structure shown in Fig. 12.6 is an acceptable process model and may be taken forward to the next step as it stands. However, it is useful — and arguably more correct — to enhance this structure further before proceeding. This involves two types of box:

- the addition of structure boxes above all selections and iterations;
- the addition of extra sequence boxes.

The latter of these two activities is not required by SSADM, but is included here as 'good practice' and leads, we believe, to a more consistent and checkable approach.

The diagram in Fig. 12.6 can be enhanced under the first activity quite readily as shown in Fig. 12.7. A decision can now be made about the need for additional sequence boxes. Sequence boxes can be added if a structure box (i.e. a box with another box below it) requires some processing to be performed. For example, the box 'process order' represents a number of operations that will need to be performed both before and after the iteration of order lines. Similarly, the iterated box 'process order line' represents some processing. To be consistent with the use of Jackson-like structures in other

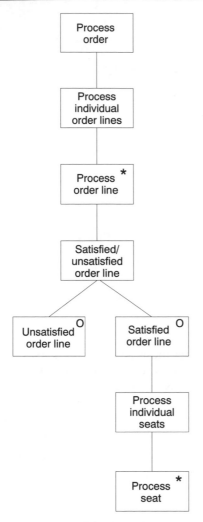

Figure 12.7 Process model enhanced to show structure boxes above selections and iterations

SSADM techniques, the 'leaves' should be the only boxes on the diagram that 'do' anything. Therefore additional boxes can be added which represent these items of processing at the appropriate points in the structure. (Alternatively, it is perfectly legitimate to show operations attached directly to structure boxes. It is the author's preference to add the extra sequence boxes.)

The expanded structure from Fig. 12.7 is shown in Fig. 12.8. This diagram shows that operations are required to set a flag to indicate that the 'order line' has been satisfied before the individual seats are processed. An extra box is not required below 'unsatisfied order line' because it is already at the bottom of the structure.

ALLOCATE OPERATIONS

The operations list compiled previously is now used to supply the operations that go with each of the bottom boxes on the structure. If a box does not have an operation associated

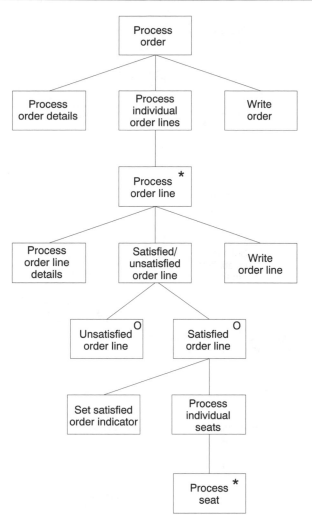

Figure 12.8 Process model with additional sequence boxes

with it, further investigation may be required to ensure that no operations are missed. If there is more than one operation, they should be added in the sequence in which they will be executed. Fig. 12.9 shows the structure derived above enhanced by the addition of operations to the bottom 'elementary' boxes. The operation numbers are taken from the list in Figure 12.5. This activity will help to check that:

- the structure is correct;
- the operations list is complete.

This is done by making sure that the structure and operations are compatible with one another. For example, if there are operations that do not fit on the structure, or if there are too many operations on a single box, additional boxes may need to be added to the structure.

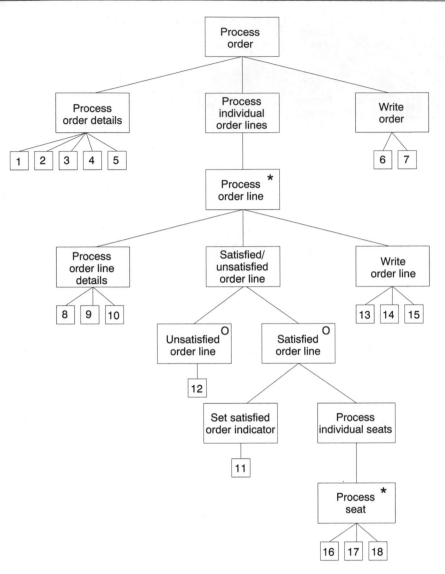

Figure 12.9 Process Model with operations added

The types of operations that are included on Update Process Models are those that appear on the Entity Life Histories plus:

Create ⟨entity⟩
Set state indicator of ⟨entity⟩ to ⟨value⟩
Write ⟨entity⟩
Read ⟨entity⟩ by key
Read next ⟨entity⟩ in set
Read next ⟨detail entity⟩ of ⟨master entity⟩ (via ⟨relationship⟩)

Read ⟨master entity⟩ from ⟨detail entity⟩ (via ⟨relationship⟩)
Invoke ⟨common process⟩
Fail if state indicator of ⟨entity⟩ outside range ⟨values⟩

ALLOCATE CONDITIONS

Each selection and iteration is further described by the use of a 'condition'. A condition states under what circumstances the selection or iteration will take place. Conditions should always be unambiguous and should be expressed in terms of something that can be tested against. For example, it is not adequate to state 'until valid' as a condition if it is not clear what constitutes 'valid'; it is better to say 'until accepted indicator = Y'. Often, a condition tests data item values or state indicator values. It may be necessary to introduce new data items into the Required System Logical Data Model so that a test can be devised for a specific condition.

There are two options for conditions under iterations: 'while' and 'until'. Roughly equivalent uses of these two types of condition would be '*while* accepted indicator = N' and '*until* accepted indicator = Y'.

There is a subtle, but important, difference between these two types of condition:

- 'while' is used only for zero-based iterations (i.e. it need not happen). In the example above, the value of the accepted indicator can be tested *before* going into the iteration and the iteration can be bypassed completely if the condition fails;
- 'until' is used only for one-based iterations (i.e. it must happen at least once). In the same example, the value of the accepted indicator is only tested *after* the iteration has been performed so can only be found to fail after one pass of the iteration.

The use of conditions is demonstrated in Fig. 12.10, which shows the completed Update Process Model for the event 'arrival of order'. In this example both the iteration conditions are 'until' as each order must have at least one line and each satisfied order line must have at least one seat.

SPECIFY INTEGRITY ERROR CONDITIONS

Here, statements are included which would cause the process to fail if the data is in an invalid state for the process to proceed. It is probable that most of these will be derived from state indicator values from the Entity Life Histories.

SPECIFY ERROR OUTPUTS

For each integrity error specified in the previous activity, error outputs are defined.

WALK THROUGH THE STRUCTURE

As a final check on the Update Process Model, it should be 'walked through', preferably with people who have not been closely involved with the construction of the model, to ensure that it makes sense as a piece of processing and meets the requirements of the event and its related functions.

Enquiry Process Models

Enquiry Process Models are produced for each enquiry in the system:

- one for each enquiry function;
- one for each fragment of enquiry used in Update Process Models.

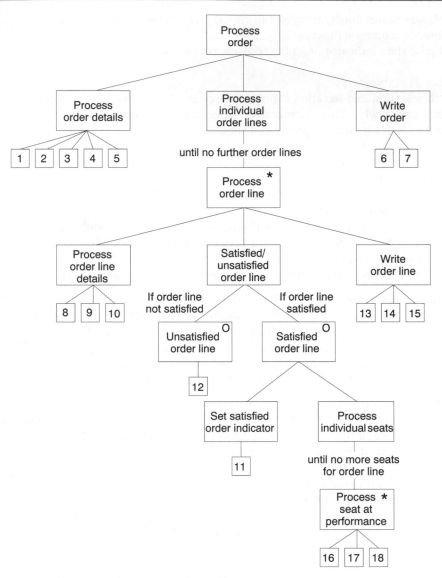

Figure 12.10 Completed Update Process Model for event 'arrival of order'

Enquiry Process Models are derived in a similar way to Update Process Models in that they are principally based on Enquiry Access Paths. However, the Enquiry Models go one step further in that the Jackson-like structure derived from an Enquiry Access Path is then merged with the output leg of the appropriate I/O Structure. The Enquiry Access Path is considered to be the 'input' to the enquiry as it shows the accesses to the stored data required for the enquiry, and the I/O Structure is considered to be the 'output' from the enquiry as it shows the structure of the data coming out of the system.

Enquiry Process Models conform to the structuring rules described in Appendix D, and are built up in a number of activities:

- specify the enquiry output;
- specify the enquiry input;
- merge input and output structures;
- add extra boxes to complete the structure;
- allocate operations to the structure;
- allocate conditions to the structure;
- specify integrity error conditions;
- specify error outputs;
- walk through the structure.

Within Enquiry Process Models the 'input' is defined as coming from the database and the 'output' is defined as the report.

These activities are demonstrated in the following sections with reference to the Opera Booking System (see Appendix F).

SPECIFY THE ENQUIRY OUTPUT

The enquiry output is specified with reference to the I/O Structure for the enquiry. Any parts of the structure that output data to the user are identified. If the I/O Structure is output-only, or the majority of the boxes are 'output', no further action is strictly necessary at this point, although the reference manuals require an output structure to be constructed.

In most cases, the structure is fairly easy to read direct from the I/O Structure. If, however, the output parts of the I/O Structure are fragmented or difficult to assimilate from the I/O Structure, it is helpful to transfer the output parts of the I/O Structure to a new diagram which is output structure only. (If this is done, the boxes should be 'hard', as this is going to form the basis of a processing structure.)

The I/O Structure and the derived output data structure for the enquiry to print tickets is shown in Fig. 12.11. In this case the transformation is not strictly necessary as the output data is clear from the I/O Structure, but it is included to demonstrate the principle.

SPECIFY THE ENQUIRY INPUT

The enquiry input is expressed in the form of a Jackson-like diagram derived directly from the relevant Enquiry Access Path. This is done in a similar way to the derivation of the initial Update Process Model from the Effect Correspondence Diagram.

The Enquiry Access Path for the 'ticket print' enquiry is shown in Fig. 12.12.

Loops are now added to this diagram around access correspondences in the same way that loops were placed round one-to-one correspondences on Effect Correspondence Diagrams. The 'looped' Enquiry Access Path for the 'ticket print' enquiry is shown in Fig. 12.13, which shows that the same rules apply as in the looping of Effect Correspondence Diagrams:

- Access correspondences are grouped together.
- Selections are looped separately.
- Iterations are looped separately.
- The 'loops' may be labelled.

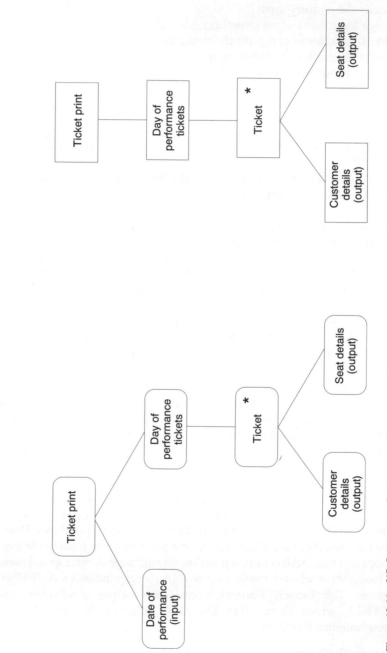

Figure 12.11 I/O Structure and output process structure for 'ticket print'

152

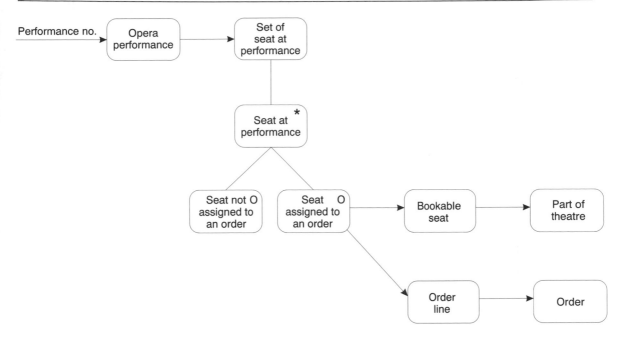

Figure 12.12 Enquiry Access Path for 'ticket print'

From this, the input structure can be derived starting from the entry point for the enquiry and working around the Enquiry Access Path as with the Effect Correspondence Diagram. The resulting process structure is shown in Fig. 12.14 together with the 'enhanced' structure derived after putting structure boxes above selections and iterations (the reasons for this are covered in the section on Update Process Models in Sec. 12.5).

MERGE INPUT AND OUTPUT STRUCTURES

The input and output structures are now merged using fairly standard techniques for merging two Jackson-like structures.

Input and output structures are placed side-by-side so that similarities and differences can be determined. In theory, the structures should be similar as the data going into the process should be of a similar structure to the data coming out of the process. In practice, the structures may be similar with slightly different terminology or there may be slight dissimilarities (structure clashes) that need to be resolved. It is less likely that there will be a major structure clash between the two structures because the I/O Structures have indirectly been used to derive the Enquiry Access Paths (the I/O Structures are used as an input to relational data analysis, which, in turn, is used to modify the Required System Logical Data Model; the Enquiry Access Paths are based on the Required System Logical Data Model).

The input and output structures for the 'ticket print' enquiry are shown together in Fig. 12.15. Although the structures look a little different, it is possible, upon examination, to see that the iteration of 'ticket' on the output structure must correspond to the option 'assigned seat' on the input structure as a single ticket will be printed for every

154

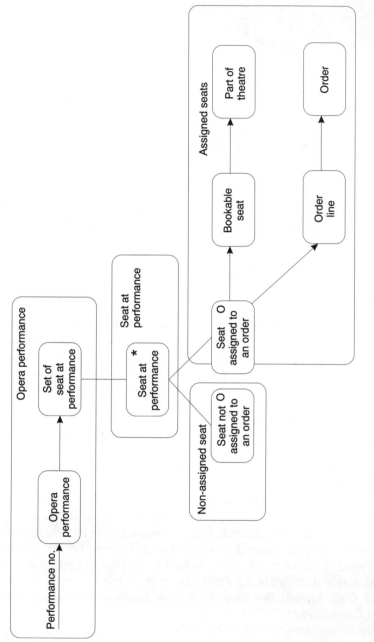

Figure 12.13 'Looped' Enquiry Access path for 'ticket print'

Initial input structure

Enhanced input structure

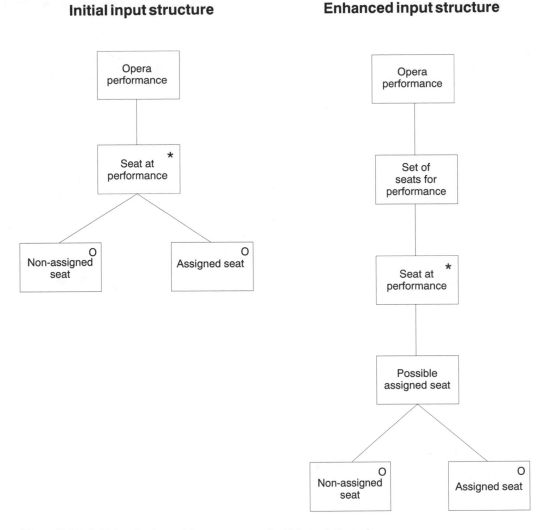

Figure 12.14 Initial and enhanced input structures for 'ticket print' enquiry

seat that is assigned. In this example, there are no structure clashes to resolve. If there are direct clashes between the structures it may be necessary to re-examine the derivation of both structures.

There are three types of clash that may need to be addressed:

- *Ordering clash*, where the sequence of components on one structure is different from the sequence on the other.
- *Boundary clash*, where the data is grouped differently in the two structures.
- *Interleaving clash*, where data items from different entities are grouped together in a way that does not correspond to the entities on the Required System Logical Data Model.

Complete input structure   **Output data structure for 'ticket print'**

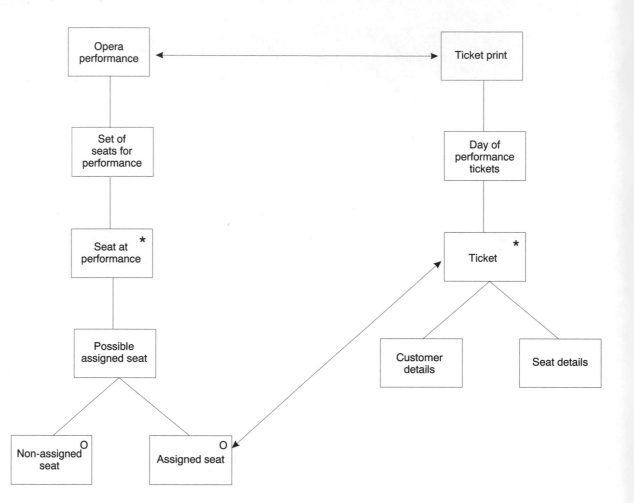

Figure 12.15 Comparison of input and output structures

The resolution of the clash may require amendments to the Required System Logical Data Model, the introduction of 'sort' processing, or the separation of processing to handle the input data from the processing for output data. Also the resolution may just involve the reordering of the input structure as no order is necessarily implied on the Enquiry Access Path.

If a major structure clash does occur, then the Required Logical Data Model should be revisited to ascertain why it does not appear to support the required enquiry.

ADD STRUCTURE AND SEQUENCE BOXES
The structure derived from directly merging the input and output structures will be an acceptable process model and may be taken forward to the next step without further

amendment. However, as for the Update Process Models, it is useful — and arguably more correct — to enhance this structure further before proceeding. This involves the possible addition of two types of box:

- Structure boxes above all selections and iterations that do not already have them.
- Extra sequence boxes to ensure all operations can be allocated to 'leaves'.

The merged structure with additional structure and sequence boxes for the 'ticket print' enquiry is shown in Fig. 12.16. This structure forms a good checkable basis for the

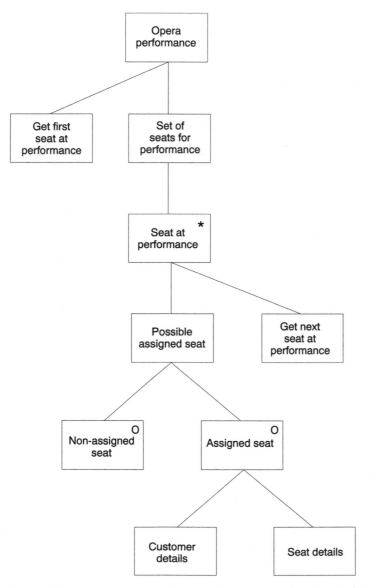

Figure 12.16 Process model for 'ticket print' enquiry showing structure boxes above selections and iterations and sequence boxes

Enquiry Process Model as only the bottom (elementary) boxes have processing in the form of operations attached and bottom boxes without operations (except null boxes) can be checked to ensure no processing has been overlooked.

ALLOCATE OPERATIONS TO THE STRUCTURE

A list of operations for each enquiry should now be compiled. The types of operation permitted for an enquiry are more limited than for the update processing as only reading and data manipulation will be performed. The types of operation used here include:

read ⟨entity⟩ by key
define ⟨set of entities⟩ matching input data
read next ⟨entity⟩ in set
read next ⟨detail entity⟩ of ⟨master entity⟩ (via ⟨relationship⟩)
read ⟨master entity⟩ of ⟨detail entity⟩ (via ⟨relationship⟩)
invoke ⟨common process⟩
fail if state indicator of ⟨entity⟩ outside range ⟨values⟩

These operations are allocated to the bottom boxes on the structure. Every bottom box should have at least one operation associated with it. If there is more than one operation, they should be added in the sequence in which they will be executed.

Figure 12.17 shows the structure derived above enhanced by the addition of operations to the bottom 'elementary' boxes. Entities that are read iteratively should be considered carefully. It is probable that an operation to read the first occurrence of the entity should be added to a box before the iteration so that the processing performed under the iteration always ends with a read of the next entity. This is what is termed 'reading ahead' and allows the processing to determine whether a record is valid before proceeding to process it.

In this example, it was not necessary to have a 'define' operation before the read of 'seat at performance', as the enquiry would want all seats belonging to the opera performance. If a subset was required, the first read would have been preceded by a 'define'.

It should be noted that it is not necessary to define an operation to deal with the 'outputting' of the report.

ALLOCATE CONDITIONS TO THE STRUCTURE

Each selection and iteration should be further described by the use of a 'condition'. A condition states under what circumstances the selection or iteration takes place. Conditions should always be unambiguous and should be expressed in terms of something that can be tested against a specific value. For example, it is not adequate to state 'until valid' as a condition if it is not clear what constitutes 'valid'.

Often, a condition tests data item values or state indicator values. It may be necessary to introduce new data items into the Required System Logical Data Model so that a test can be devised for a specific condition.

For selections, the condition defines under which circumstances that leg is to be navigated.

There are two options for conditions under iterations: 'while' and 'until'. The differences between these two types of condition are explained above in the section on

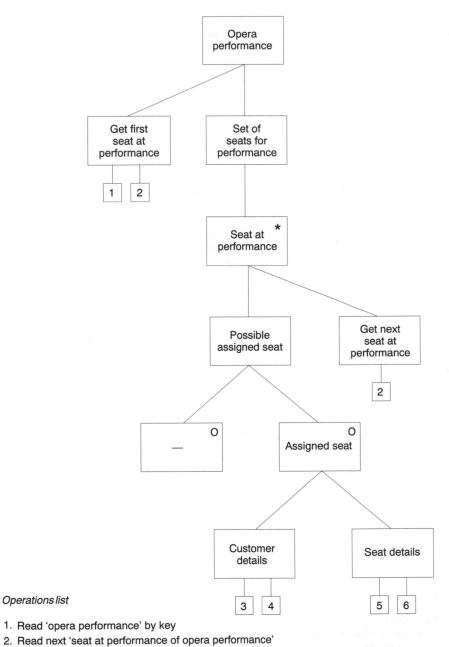

Operations list

1. Read 'opera performance' by key
2. Read next 'seat at performance of opera performance'
3. Read 'order line' of 'seat at performance'
4. Read 'order' of 'order line'
5. Read 'bookable seat' of 'seat at performance'
6. Read 'part of theatre' of 'bookable seat'

Figure 12.17 Process model with operations added

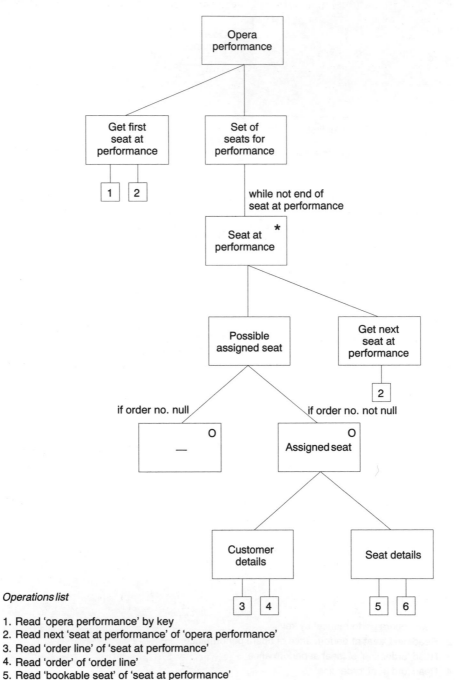

Operations list

1. Read 'opera performance' by key
2. Read next 'seat at performance' of 'opera performance'
3. Read 'order line' of 'seat at performance'
4. Read 'order' of 'order line'
5. Read 'bookable seat' of 'seat at performance'
6. Read 'part of theatre' of 'bookable seat'

Figure 12.18 Completed Enquiry Process Model for 'ticket print'

Effect Correspondence Diagrams. The completed 'ticket print' Enquiry Process Model is shown in Fig. 12.18.

SPECIFY INTEGRITY ERROR CONDITIONS

Here, statements are included which would cause the process to fail if the data is in an invalid state for the process to proceed. It is probable that most of these will be derived from state indicator values from the Entity Life Histories.

SPECIFY ERROR OUTPUTS

For each integrity error specified in the previous activity, error outputs are defined.

WALK THROUGH THE STRUCTURE

As a final check on the Enquiry Process Model, it should be 'walked through', preferably with people who have not been closely involved with the construction of the model, to ensure that it makes sense as a piece of processing and will satisfy the requirements for the enquiry.

12.6 Comparison with Version 3

The equivalent products in Version 3 to the Update Process Models and Enquiry Process Models are the Logical Update Process Outlines (LUPOs) and Logical Enquiry Process Outlines (LEPOs). Their purpose is similar in that:

- Both Process Outlines and Process Models detail the processing required to satisfy enquiries and events;
- Both Process Outlines and Process Models are based on the navigation of the data model.

The derivation of Process Outlines is more informal and less structured than the derivation of Process Models — there are no formal intermediate products between the Retrievals Catalogue of Version 3 and the Logical Enquiry Process Outlines, for example.

Process Outlines are non-diagrammatic — they list the operations required on a form. The Process Outline forms list accessing volumetrics in addition to the entities accessed and data items used in the access. The Update Process Models do not contain this type of information but they are much more structured in the way that the processing is defined and they form a better basis for subsequent design activities. Experience of using Version 4 has shown that the Effect Correspondence Diagrams and Enquiry Access Paths can be used to record the accessing volumetric information that is documented on the Process Outline forms of Version 3 if this is required.

13. Physical Design

13.1 Introduction

The purpose of Physical Design in SSADM Version 4 is to convert the Logical System Specification into a design suitable for the physical environment. The guidelines for this conversion take into account that there are factors which influence the way in which this conversion is undertaken that vary depending upon the environment of the project, namely:

- the tools and language of the implementation environment;
- the standards of the organization.

Before any design work is undertaken, therefore, a strategy for the physical design approach is developed and agreed. This strategy tailors the standard approach of SSADM to the specific environment of the project. In some cases, the entire Physical Design module may need to be restructured using, if necessary, alternative and complementary physical design techniques if the project environment demands a different approach. The guidelines for this module should be considered as a checklist rather than a set of procedures and tailoring to the environment should be undertaken at the outset.

Thus the Physical Design Module is composed of two main parts:

- Preparing for Physical Design, which involves the development of a Physical Design Strategy.
- Developing data and process designs to the standards defined in the Physical Design Strategy.

The output of the Physical Design Module is the Physical System Specification, which is based on the Logical System Specification and consists of:

- the Physical Data Design;
- the Physical Process Specification.

Physical Data Design is undertaken in a series of activities, working from an initial physical data model, which is produced by applying general rules, through a product-specific version of the model to a process of optimization where the model is tested against performance and sizing objectives.

Physical Process Specification is based on the development of a Function Component Implementation Map (FCIM), which describes a framework relating the physical 'fragments' of a function to one another and to the corresponding areas of the Logical

System Specification. The FCIM also helps to identify common and duplicate elements and to plan the implementation of components. The Function Component Implementation Map can be expanded to show the role of a Process Data Interface (PDI) if required. The Process Data Interface is assumed to be a set of facilities that control the access of data by the programs. This approach can make the design of data relatively independent of the design of processing.

Throughout this chapter we have assumed that a Function Component Implementation Map is produced for each function. It should be noted, however, that if a more sensible unit of physical design is identified, then an FCIM should be developed for this unit instead of for each function.

Although a number of the logical products can have an impact on the Physical Design, there are some basic dependencies that can be defined. These are shown in Fig. 13.1.

The precise procedures for Physical Design are different for each project. In general, however, an underlying assumption of the SSADM Physical Design procedures is that the same sort of activities need to undertaken whatever the environment. These activities are not always undertaken on paper. There are no explicit assumptions about the point at which implementation could be initiated and a number of the steps could be more appropriately addressed using the tools and dictionary of the target system. For example, if the development software has tools to help optimize data designs, it is sensible to implement an unoptimized data design and to use the system tools to help in the subsequent optimization.

13.2 Naming conventions

The *Physical Design Strategy (PDS)* is developed at the outset of Physical Design and defines the procedures and standards to be adopted.

Installation Development Standards are an input to the Physical Design Strategy. *Application Development Standards* are identified or developed during the construction of the Physical Design Strategy.

The *Physical Environment Specification* is an overview of the physical system and is based on the Technical Environment Description developed during Technical System Options and the Requirements Specification.

A *Physical Environment Classification* is developed to describe the target implementation environment in terms of the facilities for data handling, performance and process development using the following:

- *Processing System Classification*;
- *DBMS Data Storage Classification*;
- *DBMS Performance Classification*.

The *Physical System Specification* is based on the Logical System Specification and includes the following:

- *Physical Data Design*;
- *Physical Process Specification*.

The *Physical Design* is composed of the *Physical System Specification* combined with supplementary physical implementation information.

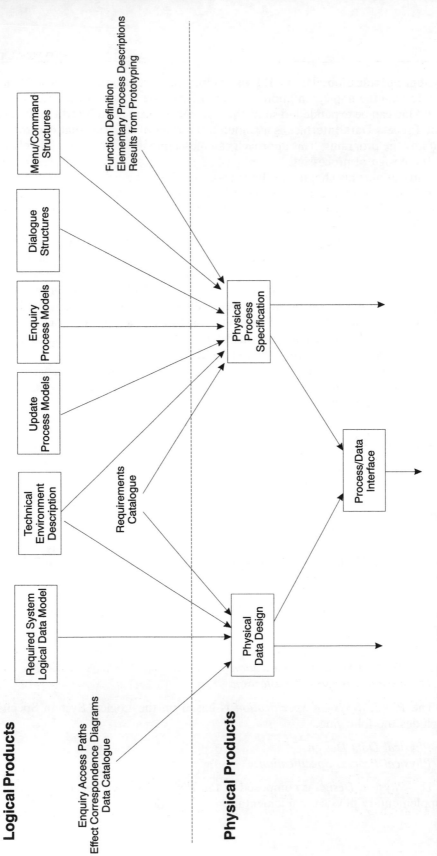

Figure 13.1 Mapping of Logical to Physical Design products

Physical Data Design is concerned with physical data placement and DBMS optimization.

Physical Process Specification involves the production of a *Function Component Implementation Map*. A FCIM is produced for each physical function showing how all the components of the function are grouped together.

The *Universal Function Model* may be used as the basis for a *Specific Function Model* if part of the processing is to be implemented procedurally.

Success units, which are identified during the construction of the Function Component Implementation Map, are defined as sets of processing that each either succeed or fail as a whole. The Function Component Implementation Map can also show the role of the Process Data Interface (PDI).

13.3 Place in structure

The Physical Design Module encompasses Stage 6: Physical Design. All Physical Design activities, except those that may be undertaken to support Technical System Options, are described as part of the Physical Design Module. Physical Design consists of the following steps:

610 Prepare for Physical Design
620 Create Physical Data Design
630 Create Function Component Implementation Map
640 Optimize Physical Data Design
650 Complete Function Specification
660 Consolidate Process Data Interface
670 Assemble Physical Design

It is expected that the steps undertaken will depend upon the environment:

- If an application generator, or fourth-generation language (4GL) is used, it is expected that the generation of the system can be initiated after Step 630. If no procedural code is required, Step 650 is optional.
- If a third-generation language (3GL) is used, it is expected that coding will be initiated immediately after the end of Step 670.

13.4 Place in Product Breakdown Structure

The Product Breakdown Structure for Physical Design can be found in Appendix B, which gives the complete Product Breakdown Structure for all modules of SSADM. All the Physical Design Products are products of the Physical Design Module and there is no overlap with other modules.

13.5 Notation and use

The first step of Physical Design, the preparation for Physical Design, can start before the completion of Stages 4 and 5, provided that the implementation environment and/or local standards are known, which is often the case where a new application is added to an existing computer system. This step can be thought of as equivalent in purpose to Step 110, which starts the project. At Step 110, not enough information is available to plan Physical Design in detail. Thus, at Step 610, a number of the activities of Step 110 are repeated:

- A Product Breakdown Structure and Product Descriptions are developed for Physical Design products.
- An Activity Network and Activity Descriptions are developed to plan the tasks to be undertaken.

The Physical Design Strategy is developed as part of this first step. The Physical Design Strategy defines how the Logical System Specification is to be mapped over to a specific implementation environment. This is based on schemes for classifying the processing, performance, and data-handling characteristics of the target environment. These are:

- The Processing System Classification, which provides a guide to the features of the development software and includes factors such as the types of tools provided, whether the language(s) are procedural or non-procedural, error handling mechanisms, and dialogue handling.
- The DBMS Data Storage Classification, which defines how data is stored and retrieved by the target DBMS and includes factors such as the relationship representation, amalgamation of entity and relationship data, key representation in relationships, retrieval mechanisms and restrictions.
- The DBMS Performance Classification, which defines the data access and update properties in order to assess the performance of particular functions and includes factors such as transaction logging, recovery logging, space management, and standard timing factors such as disk access time.

The Physical Design of data and processing are closely interrelated. The data design needs to support the processing and the processing needs to be built around the structure of the data. In overview, the approach to physical data and process design is as follows:

- The data design is addressed first. The Required System Logical Data Model is converted into an unoptimized physical version in two stages, the first following general rules to convert the data model into a universal model, which then has product-specific rules applied.
- The functions are addressed by constructing Function Component Implementation Maps which define the physical fragments of each process and their relationship to components of the Logical Design.
- The data design may then be optimized with reference to the Function Component Implementation Maps and the performance and data objectives documented on Function Definitions and the Requirements Catalogue.
- The Function Definitions can then be completed for systems that are to use procedural languages for some or all of the programs (for non-procedural implementations, the Function Component Implementation Map is considered of sufficient detail at this point).
- The interface between the processing and the database is defined and the physical design can then be assembled.

The following sections describe in more detail the approaches to Physical Data Design and Physical Process Specification.

Physical Data Design

Physical Data Design is undertaken in a number of activities:

- preparation;
- create Physical Data Design;
- meeting objectives;
- complete Physical Data Design.

PREPARATION

The preparation for Physical Data Design is part of the initial preparation activity that produces the Physical Design Strategy. The following are activities particularly associated with Physical Data Design:

- Studying the implementation environment to understand the DBMS or file handler characteristics (this is likely to require the involvement of 'experts' in the chosen environment).
- Classifying the DBMS in terms of data storage and performance characteristics;
- Creating the Physical Design Strategy for data design.

CREATE PHYSICAL DATA DESIGN

The initial creation of the Data Design produces a design that could be implemented on the target environment but is not optimized in any way. As such, it may be regarded as a 'first cut' design.

The conversion of the Required System Logical Data Model to its physical equivalent is undertaken in a series of activities, the first seven of which are general to all DBMSs and the last one of which has product-specific rules applied. The resulting design is an implementable design that will not necessarily be able to satisfy space and performance objectives.

The eight activities of Physical Data Design are:

1. Identify features of Required System Logical Data Model required.
2. Identify required entry points.
3. Identify roots of physical groups.
4. Identify allowable physical groups for each non-root.
5. Apply least-dependent occurrence rule.
6. Determine block size to be used.
7. Split physical groups.
8. Apply product-specific rules to design.

A number of these activities are illustrated here by reference to the Opera Booking example. The Required System Logical Data Structure for the Opera Booking System is shown in Fig. 13.2 for reference in the following sections.

Other SSADM documentation used in these activities are:

- Enquiry Access Paths;
- Effect Correspondence Diagrams;

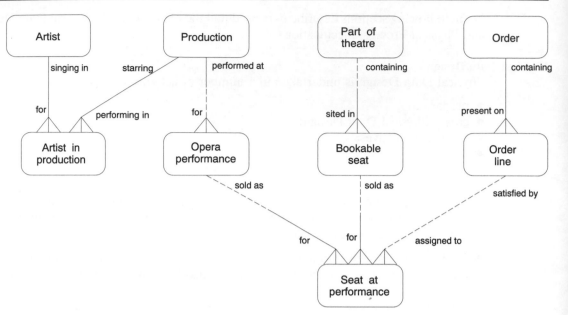

Figure 13.2   Required System Logical Data Structure for the Opera Booking System

- Data Catalogue;
- Entity Descriptions.

1. Identify features of Required System Logical Data Model required
The Required System Logical Data Model contains two types of information:

- data and relationships that are of relevance to the Physical Data Design;
- business rules and constraints that have no direct mapping to the Physical Data Design.

In order to construct a Physical Data Structure, it is necessary first to determine which elements of the Logical Data Structure are of relevance. Table 13.1 shows the components of a Logical Data Structure and their mapping to a Physical Data Structure.

Given that only some elements of the Logical Data Structure are relevant to the derivation of a Physical Data Structure, it is useful to redraw the Required System Logical Data Model so that it shows only those items of relevance. In addition, it is useful to annotate the diagram with some volumetric information.

The resulting diagram from the Opera Booking System example is shown in Fig. 13.3. Here, the relationship representation indicates the following:

- solid lines are mandatory from master to detail;
- lines with circles show that the relationship is optional.

The numbers added to the entities and relationships show the average predicted occurrences for each entity, and the average number of relationship occurrences for each master occurrence. In addition, the boxes have changed from the 'soft' type to the 'hard' type to show that the entities relate directly to physical files.

Table 13.1 Mapping of Logical Data Structure to Physical

Logical Data Structure	*Physical Data Structure?*
Entities	Direct mapping to files, etc.
Relationships	
Mandatory from detail to master	Needed to determine grouping of entities
Mandatory from master to detail	Treated as optional
Totally mandatory	Treated the same as mandatory from detail to master
Totally optional	Treated as optional
Exclusive relationships from master to several details	Not relevant
Exclusive relationships from detail to several masters	All relationships in set treated as optional
Partial models	Not relevant — total design should be used
Relationship labels	Not relevant

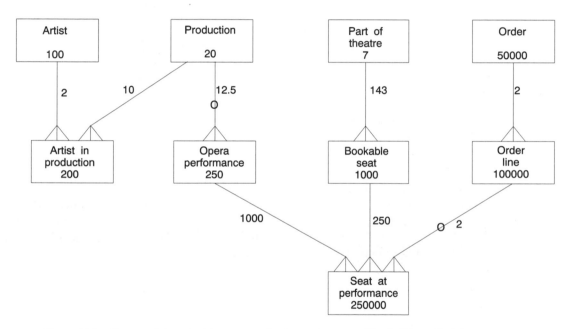

Figure 13.3 Converted data model to remove irrelevant features of Logical Data Model

2. Identify required entry points

An entry point is an entity that is accessed directly as part of some update or enquiry processing by means of the primary key or some other combination of data items. By examining all Effect Correspondence Diagrams and Enquiry Access Paths, it is possible to determine which entities are to be used as entry points into the data model. For example, on the Enquiry Access Path for 'list all seats for a specified part of theatre' (Fig. 11.7), the entry point is the entity 'part of theatre', which is accessed by its prime key 'theatre area'. Also, the enquiry to list all artists appearing on a specified date

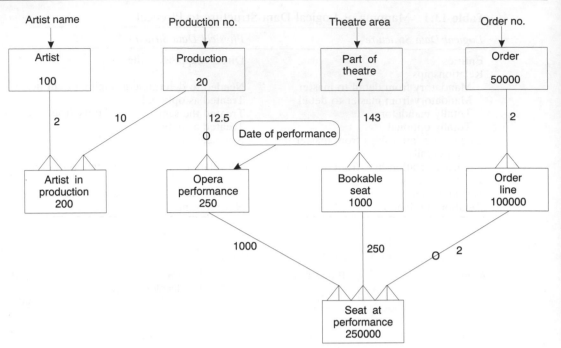

Figure 13.4 Physical data structure with entry points shown

(Fig. 11.8) has as an entry point the entity 'opera performance' that is accessed by the data item 'date of performance'.

The entry points are all marked on the physical data structure as shown in Fig. 13.4. All accesses by items other than the primary key of the entity are circled to denote the need for a secondary index. Therefore the item 'date of performance' is circled to show that it is not the primary key of 'opera performance'.

3. Identify roots of physical groups
A root entity is identified by applying one of the following criteria:

- It is an entity without a master;
- It is an entry point.

The only exception to this is where the entity is an entry point and has a composite key, one part of which is the key to an entity already identified as a root. Under these circumstances the entity is not marked as a root.

Root entities are denoted by a line across the top of the box as shown in Fig. 13.5. As can be seen here, 'opera performance' is a root entity even though it is the detail of a 'root' entity because the key 'performance number' does not contain the key of its master, which is 'production number'.

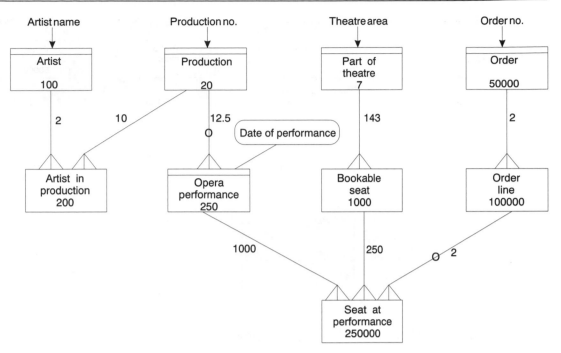

Figure 13.5 Physical data model showing root entities

4. Identify allowable physical groups for each non-root

All entities that have not been identified as roots need to be allocated to a root entity to form physical groups. The allocation of non-root entities to root entities is based on the following criteria:

- An entity is placed in the same group as a *mandatory* master (optional relationships are ignored).
- If there is a choice, the entity should be placed with the master whose key is part of its own.

Under these circumstances 'order line' is grouped with 'order' and 'bookable seat' with 'part of theatre'.

Link boxes are likely to be allocated to two different groups at this stage, but this will be resolved in the next activity. For example, the entity 'seat at performance' is allocated to 'bookable seat' and 'opera performance' by applying this rule.

5. Apply least-dependent occurrence rule

If an entity has been allocated to more than one group in the previous activity, this is resolved by examining the relationship occurrences. It should be placed with the entity where the relationship occurrences are least in relation to the root entity of the group. In the opera booking example, the entity 'seat at performance' is placed with 'opera performance' because there are only 1000 occurrences of the relationship per occurrence

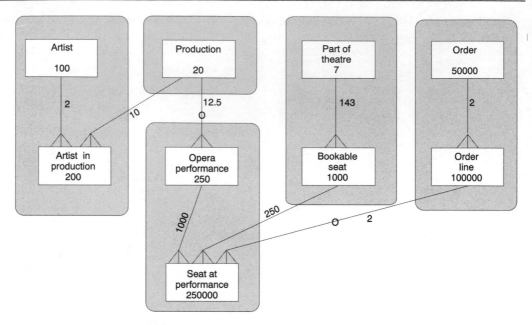

Figure 13.6 Physical data design with physical groups shown

of 'opera performance' as opposed to 35 750 (143 × 250) occurrences for each occurrence of 'part of theatre'.

The resulting physical data design is shown in Fig. 13.6.

6. Determine block size to be used
In many cases, the block size will be determined by the selected DBMS. If not, the designer will select the most efficient block size, which is a compromise between wanting to maximize the amount of data handled together while trying to minimize the buffer sizes required to transfer the blocks.

7. Split physical groups
The sizes of the physical groups identified in activity 5 are now compared to the chosen block size. The size of each entity can be calculated with reference to the Entity Descriptions and the Data Catalogue.

If a physical group does not fit the chosen block size, the group should be split so that each group is less than or equal to the block size.

8. Apply product-specific rules to design
The physical data design is now mapped onto the physical facilities provided by the chosen DBMS. This is assisted by referring to the DBMS Data Storage Classification. As a result, a data design is produced that can be implemented on the target DBMS.

It is worth noting that a large number of DBMSs allow the direct implementation of a Logical Data Model without the need to specify physical groups. In this case, only activity 1 may be required before converting the design into its implementation-specific equivalent.

MEETING OBJECTIVES

The design produced by the application of the eight activities may need to be modified to meet objectives described in the Requirements Catalogue and Function Definitions. There are two main categories of objective, namely *space* and *performance*.

The data design should be tested to ensure that all objectives are satisfied. This may be done using a paper-based method or by using optimization tools supplied with the Database Management System.

Optimization is a trade-off between two factors:

- The performance and space requirements should be met to make the system run as efficiently as possible.
- The data structure should remain as close to the Required System Logical Data Model as possible to support easy maintenance, flexibility, and comprehension of the design.

The process of optimization is iterative in that it may need to be repeated and tested a number of times before the objectives are met. The basic activites of optimization are as follows:

- set storage and timing constraints;
- estimate storage requirements;
- restructure design to fit storage constraints;
- estimate resource times of major functions;
- identify problem areas.

These tasks require a deep knowledge of the implementation environment or sophisticated tools to support them. The following sections give a brief overview of the areas addressed.

Set storage and timing constraints
Storage and timing constraints should be expressed as service level requirements in the Requirements Catalogue and Function Definitions.

Estimate storage requirements
A rough guide to the size of each physical group is gained by multiplying the size of each entity within that group by the number of occurrences. This needs to be enhanced to take account of indexes, pointers, etc. There may also be additional space required for security purposes.

Restructure design to fit storage constraints
If the storage objectives are not met, there are a number of ways of reducing the size of the data. This may depend upon the facilities of the DBMS used. There are also ways of manipulating the data, such as using codes instead of full names, removing historical data, or removing redundancy.

Estimate resource times of major functions

It should be possible to identify functions that are likely to be critical in determining whether the system will perform as it should. These are either very high-volume functions or functions that access a large number of entity occurrences. These functions should be compared to the Physical Data Design to determine which physical data groups will be accessed. By using the DBMS Data Storage Classification and DBMS Performance Classification forms it should be possible to estimate the time taken to complete each of the critical functions.

Identify problem areas

Once the timings for each critical function have been estimated, they can be compared with the objectives stated on the Function Definitions and the Requirements Catalogue. If the objectives are not met, it will be necessary to decide what should be done. There are a number of things that may be done to the data structure to improve access times, e.g. data redundancy can be introduced or calculated fields can be held. Alternative DBMS facilities might help to improve performance. It is also possible that the performance objectives can be adjusted if the discrepancy is small.

If an adjustment is made, the timing calculations should be repeated and further adjustments made until the objectives are met.

COMPLETE PHYSICAL DATA DESIGN

The completion of the Physical Data Design is undertaken after the optimization exercise has been completed. It consists of the following tasks:

- validation of impact of imposed sequencing;
- documentation of sorting requirements;
- identification of processing optimization requirements;
- updating of the service level requirements on Function Definitions and Requirements Catalogue;
- recording of the optimization decisions;
- using simulation to validate the performance of the final design.

Physical Process Specification

Physical Process Specification deals with the conversion of the products of the Logical Design into programs, physical I/O formats, and physical dialogue designs.

Physical Process Specification is undertaken in a number of activities:

- prepare for Physical Design;
- create Function Component Implementation Map;
- complete Function Specification;
- consolidate Process–Data Interface.

These activities are described in the following sections.

PREPARE FOR PHYSICAL DESIGN

The development of the Physical Design Strategy involves mapping the features of the Logical Design onto the facilities offered by the development environment in order to plan how to proceed for the remainder of Physical Design. As part of this, the

Application Development Standards are defined. These identify criteria for choosing implementation paths for components of the functions, specifying how to apply those criteria to different product features, customizing the procedures for Physical Design to incorporate product-specific guidelines, and formulating program specification standards.

The tasks involved in the preparation for Physical Design that are specific to the Physical Process Specification are as follows:

- Studying the implementation environment to understand the best practices and approaches. This may require consultation with experts in the particular hardware and software chosen.
- Producing the Processing System Classification to identify the various facilities that can be used for implementing processing (such as whether the languages are procedural or non-procedural, types of tool provided, and dialogue handling).
- Specifying the Application Naming Standards for use in the production of documentation to incorporate any limitations imposed by the physical environment or by installation standards.
- Developing the Physical Design Strategy to identify the criteria for choosing how to implement particular components of functions, specify the activities required, and design the best means of building the Function Component Implementation Map.

The resulting strategy defines how the physical processes (modules) will be implemented (e.g. by procedural language, by non-procedural language, or more commonly by a mixture).

CREATE FUNCTION COMPONENT IMPLEMENTATION MAP

A Function Component Implementation Map (FCIM) is developed for each function or its physical equivalent. Its precise form is decided as part of the Physical Design Strategy. Whatever the precise method chosen for documenting the FCIM, however, its purpose is the same, namely to pull together all the different components that will implement a function and show how they fit together. It is not so much a map as a 'paper clip'!

In addition, a Function Component Implementation Map should show how each part of the Logical Design maps onto the physical components. Thus, it will contain:

- dialogues;
- physical screens;
- operations;
- events;
- enquiries;
- groups of entities;
- error handling;
- control handling;

and so on.

A Function Component Implementation Map can be documented simply as a list of the parts (or 'fragments') of the function, indicating cross-references. Alternatively, it may be considered useful to document the interrelationships between elements in the form of a 'network diagram', showing how all the parts fit together.

To construct the Function Component Implementation Map, the areas of processing that are 'physical' and have not been addressed so far are defined. For this step a number of inputs can be used to provide the information necessary. These include:

- Requirements Catalogue, for non-functional requirements;
- Function Definitions, for service level requirements;
- Elementary Process Descriptions.

The Elementary Process Descriptions may be useful as a description of the processing required; however, they are not likely to provide a major input as:

- the requirements may have been refined since the Elementary Process Descriptions were produced;
- the processing defined by the Elementary Process Descriptions should already be encompassed by the Logical Design products Update Process Models, Enquiry Process Models, and Dialogue Designs.

Seven activities are undertaken to define Function Component Implementation Maps:

- remove duplication;
- identify common processing;
- define success units;
- specify syntax error handling;
- specify controls and control errors;
- specify physical I/O formats;
- specify physical dialogues.

The first two activities are done across a number of functions and the remaining five for a specific function. The resulting products should be documented, with cross-references, and related back to the relevant elements of the Logical Design for traceability.

The Physical Design Strategy should identify which of these activities are relevant to a specific project. It should also specify how they will be done. These activities are described in more detail in the following sections.

Remove duplication
The purpose of this activity, together with the second activity, is to identify areas of processing that are the same across a number of different functions so that they only need to be specified once and thereafter reused wherever they are required.

The starting point is to look at the Update Process Models for the events that make up the update functions:

- any Update Process Models that are initiated by retrieval of data are split into 'enquiry' and 'update' fragments;
- similarly, any validation or initial checking processes are separated from the main update part of the structure.

Identify common processing
Common processing may already have been identified and documented using the techniques of SSADM with suitable cross-references. At this point, however, it may be

possible to identify additional areas of common processing by examining the Update and Enquiry Process Models.

The fragments identified in the first activity should be compared. It is possible that the initial retrievals associated with updates have already been separately specified as enquiry functions. It is very likely that the same validation processing appears in a number of places. Any such common pieces of processing should be separately identified as common routines.

Additionally, common processing might be identified where:

- events are shared between more than one function (this is especially likely to occur when two functions have been put together to form a third function (or 'super-function') because of the users' requirements for grouping functions together);
- sets of operations or parts of the structure are the same across a number of events;
- similar sets of entities are accessed;
- calculation or conversion routines are used by a number of different events. (In fact it may be useful always to define calculation and conversion routines as common as they are liable to be reused in the future even if not now.)

Common processing can be identified at a number of levels:

- whole functions which may be incorporated into 'super-functions';
- events which appear in more than one function;
- sets of operations which are reflected in the Update and Enquiry Process Models.

Define success units

A 'success unit' is defined as a set of processing that either succeeds or fails as a whole. Within a single function, there may be a number of success units, each of which has a particular objective. Success units are the basis for defining the modules of processing and may be the basis for deciding upon database updating policy. Success units may equate to events or groups of related effects occurring within the same function invocation.

Success units are generally only relevant for update functions, though some of the more important, complex enquiries may also be split down in this way.

Specify syntax error handling

This involves the identification of the validation routines associated with the input of data, specifically at screen level. There are two main types of validation:

- *syntactic*, which can be applied without access to other data values, e.g. length of fields and data types;
- *semantic*, which usually compares attributes with definitions or ranges defined centrally.

The semantic validation, which includes the checking of state indicator values, should already have been addressed in the derivation of the Update and Enquiry Process Models.

The precise definition of syntax validation and error handling is best left until Physical Design when the target environment is known as there are many different ways of specifying it. For example:

- If the implementation method handles dialogues from within the central application programs, then the validation will be embedded in the code for the application.
- If the implementation environment has an active data dictionary, the syntax validation might be held with the data item definition.
- If there are advanced screen-handling facilities, the validation may be defined at screen level.

Specify controls and control errors

The Requirements Catalogue should include a definition of the types of input and update controls required. During Physical Design, these requirements stated generally in the Requirements Catalogue need to be applied to the relevant functions.

Types of control that may need to be considered are:

- controls to reduce data errors such as check digits and hash totals;
- controls to ensure the correct use of the system functions, e.g. by the introduction of a security system to limit the use of certain functions.

Specify physical I/O formats

Physical I/O formats are the implementation-specific definitions of: screens and reports. Each function has input and output data defined as I/O Structures and Dialogue Structures. These logical definitions define, in detail, the data items that will be passed to and from the processing. They need to be enhanced to include the extra information that the physical system will require.

Examples of the type of additional information that may be required are:

- fields to handle the controls identified above;
- error messages for syntax and semantic errors;
- file transfer mechanisms;
- system messages;
- report and screen layouts.

For off-line functions, the I/O Structures are used as an input to the design of data streams and outputs. For on-line systems, the Application Style Guide should be consulted to define the layout requirements for screens as an input to the design of the physical dialogues in the following activity.

Specify physical dialogues

Physical dialogues define the actual interaction between users and the system and are needed to map the Dialogue Structures developed during Logical Design onto the facilities and features of the target system.

The Dialogue Structures produced during Logical Design provide the logical group-ings of data in the form of 'logical groupings of dialogue elements' (LGDEs). These are used as the basis of the Physical Dialogues together with the Physical Design Strategy, the Menus and Command Structures, the results of prototyping and the Application Style Guide.

Some of the issues that will be addressed here are:

- screen design (based on Application Style Guide and universal guidelines for screen design);
- physical data group design (based on the Logical Groupings of Dialogue Elements);
- physical dialogue design (based on Physical Design Strategy and facilities of the system for handling the triggering of success units);
- dialogue components to handle errors (based on syntactic errors, control errors, and semantic errors, possibly utilizing system facilities for standard error messages);
- dialogue components to handle enquiries and searches (based on the common enquiry elements identified above);
- dialogue navigation (based on Dialogue Control Tables);
- implementation of Menu and Command Structures (based on the Menu and Command Structures, the results of specification prototyping and the Application Style Guide).

In addition, a full definition of the type of help that will be provided should be produced, based on the Dialogue Level Help described in Logical Design.

COMPLETE FUNCTION SPECIFICATION

Whether the processing for the function is defined further or not depends upon whether it will be implemented procedurally or non-procedurally:

- Procedural languages require a more detailed specification of how the processing will be performed. The use of these requires further specification.
- Non-procedural languages require a broad specification of what is to be achieved without stating how it will be achieved. The use of these should not require further specification once the Function Component Implementation Map has been completed.

For the areas of the system that are developed procedurally, modules should be defined and full program specifications produced. These program specifications are constructed to conform with the project standards defined or identified within the Physical Design Strategy. Also, the Function Definitions may be updated to reflect the design decisions made.

The issues that may be addressed include:

- production of a specific function model showing individual processes and linking data streams;
- the use of program design methods;
- design of input/output subsystems;
- combining elements of functions into run units.

CONSOLIDATE PROCESS–DATA INTERFACE

The definition of a Process–Data Interface is an optional activity within Physical Design. It can be a useful concept in many environments by helping to maintain a static view of the data from the processing's viewpoint, even when the actual structure of the data changes.

A Process–Data Interface is effectively a 'filter' between the processing and the data. It contains standard routines that can accept data access requests from the processing based upon one view of the data and ensure that the correct data is accessed and passed back to the process from the physical data structure. In theory, the use of a Process–Data Interface should allow the design of processing to be based purely on the Required System Logical Data Model, which has the benefit of understandable specifications and maintainable designs.

If a Process–Data Interface is to be utilized, it is necessary to identify the functionality that is supported by the interface and to determine what is required in the specification of programs to best utilize the interface. The implementation environment may provide facilities for implementing the Process–Data Interface, or it may need to be implemented by the use of standard routines.

It is very important to know exactly what a Process–Data Interface is going to be able to achieve to avoid 'gaps' in the specification of processing. In many cases, the programs may need explicitly to invoke the Process–Data Interface routines, and in others, the use of the interface may be transparent.

13.6 Comparison with Version 3

The Physical Design procedures have been completely rewritten for Version 4. Explicit tasks have been introduced to encourage developers to define their own procedures for Physical Design based on the tools available in the implementation environment and installation standards.

There has been a substantial shift in emphasis away from a 'procedural' Stage 6 to a 'non-procedural' approach. Apart from the eight activities of Data Design, there are very few 'techniques' described. Instead, there are a number of comprehensive guide-lines and checklists for consideration.

Another major difference is that only Data Design and Physical Process Specification are addressed in Version 4, whereas Version 3 also contained steps that addressed plans, testing, documentation, etc. These issues are now addressed by Project Procedures, which are considered to be outside the scope of 'core' SSADM.

Having pointed out the differences, however, there are some significant similarities between the two versions in that the overall approach to Physical Design is very similar:

- first cut data design;
- overall specification of functional areas;
- data design optimization;
- completion of program specifications.

This is roughly an overview of both versions' approach to Physical Design. The differences are that Version 3 contained the 'what' and the 'how', whereas Version 4 concentrates on the 'what' with a multitude of suggested approaches for the 'how' without defining any specific techniques.

14. Project Procedures

14.1 Introduction

SSADM provides the procedures and techniques for systems analysis and design within a project. These procedures and techniques fit within the complete infrastructure for a project and are supplemented and controlled by complementary activities. These complementary activities are just as important as the SSADM activities but they are of a different nature to the SSADM activities and are therefore not included in the structure and tasks of SSADM.

The complementary activities of a project are termed 'project procedures'. They are described briefly within SSADM in order to:

- describe their influences upon the practice of SSADM;
- show where in the SSADM Structure they are of relevance;
- understand the context of SSADM within the project as a whole.

The following project procedures are described within SSADM:

- project management;
- quality assurance;
- risk assessment and management;
- capacity planning;
- testing;
- training;
- take-on;
- (technical) authoring;
- standards.

The definition of the project procedures within SSADM shows how the method interfaces with the individual techniques and should not be considered a full description of each of them.

14.2 Naming conventions

The activities that complement SSADM within a project are called *project procedures*. The interfaces between SSADM and project procedures are controlled by means of an *information highway*.

14.3   Place in structure

A number of the project procedures provide direct input to the steps and stages of SSADM and receive outputs from SSADM. These interface with the structure of SSADM at predefined points. These interfaces are described in Table 14.1. In the table those stages in parentheses are points where the project procedure has a minor influence, whereas those stages not in parentheses have a formal interface to the project procedure.

Other project procedures are undertaken in parallel with the SSADM stages and steps. They influence, and are influenced by, the SSADM activities but do not have specific inputs and outputs. These procedures are summarized in Table 14.2.

The remaining project procedures provide a framework within which SSADM activities are conducted. The interfaces are of a general nature. The project procedures in this category are:

- *project management*, which defines the procedures surrounding SSADM;
- *standards*, which might affect the way SSADM activities are conducted at all stages.

Table 14.1   Specific interfaces between project procedures and SSADM stages

Project procedure	Stages	Type of interface
Risk assessment	(2),4,(6)	The options require the assessment of risks as part of the option description
Capacity planning	(2),4	Capacity planning techniques are used in defining and checking the capacity required of a new system as an input principally to Technical System Options
Testing	4,6	Test criteria are developed based on the Requirements Catalogue from Technical System Options through to Physical Design

Table 14.2   General interfaces between project procedures and SSADM structure

Project procedure	Relationship with SSADM steps and stages
Training	Training requirements will be identified throughout the project. At project initiation, both developers and users require training. Further training needs in specialized areas may be identified at a number of different points in all subsequent stages
Take-on	The sources of data for take-on are identified during the investigation of the current environment. The take-on strategy and specification is developed in parallel with the remaining steps and stages
Authoring	A number of the SSADM products require technical authoring skills. In addition, the design and production of the system documentation (including user manuals and operations manuals) may be initiated during Technical System Options and continue through Physical Design
Quality assurance	Procedures for quality assurance encompass the whole project. Specific interfaces are at the end of each SSADM stage but the process of assuring quality affects the whole method

14.4 Place in Product Breakdown Structure

The Product Breakdown Structure defines all the products of SSADM and their interrelationships. Project procedures do not formally appear within the Product Breakdown Structure, rather they either make a direct contribution to the construction of some of these products or define the way in which they are produced. This influence cannot be described in detail as it depends upon the content of the individual project procedures.

14.5 Project procedures and SSADM

Project procedures are activities that complement SSADM tasks and procedures. They are therefore not part of SSADM. As such, they are not described in as much detail as the central activities of SSADM. The project procedures described within SSADM are not a complete set but act as examples of the complementary activities that are often undertaken and need to be planned for.

The interface between the project procedures and SSADM is controlled by means of an information highway. This defines what can be passed to and from SSADM activities and what quality criteria can be applied.

The project procedures described here are:

- project management;
- quality assurance;
- risk assessment and management;
- capacity planning;
- testing;
- training;
- take-on;
- (technical) authoring;
- standards.

Project management

Strong project management is of vital importance to the success of a project — SSADM alone does not ensure that the objectives of the project are met. Project management needs to ensure that SSADM is directed properly and that products are delivered when they should be and to the correct standard of quality.

Project Management covers a number of topics, including:

- organization;
- control;
- planning;
- estimating.

It is advised that a project management method, such as PRINCE, should be used in conjunction with SSADM to provide these constituents. The project management method should be thought of as complementary to SSADM.

ORGANIZATION

The organization of a project defines the roles of the participants and their specific responsibilities. These may vary depending upon the method or style of project management adopted. Typical roles within a project are represented in Fig. 14.1.

The responsibilities of these roles might be as follows:

- The *project board* has overall control and responsibility for the project. It has the authority to commit resources and makes decisions about the various options through the project.
- The *project manager* has direct responsibility for managing the project. He or she is responsible to the board for the project plans.
- The *stage/module manager* has the technical responsibility for a part of the project. He or she is responsible for ensuring the production of quality products within budget and on time. On many projects the stage/module manager and the project manager are the same person.
- The *development team* consists of analysts, designers, and specialists brought in for specific skill requirements.
- The *project assurance team* is responsible for ensuring the quality of the products of SSADM and the continuity of the project.

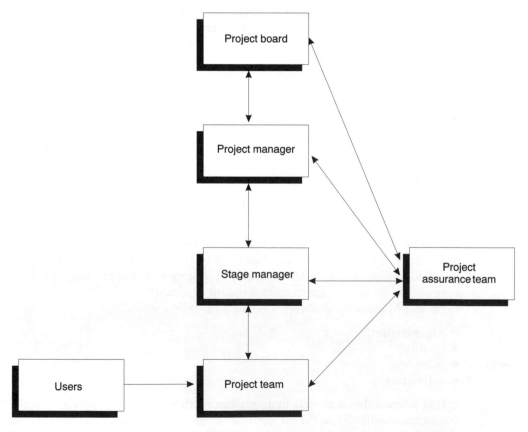

Figure 14.1 Typical organization of a project

- The *users* are the source of information on the current environment and requirements for the new system. They will be represented on the project board and project assurance team for decision making.

CONTROL

Project control compares the achievements of a project with the planned achievements and takes any action necessary to deal with discrepancies. Control is exercised at different levels of the management structure depending upon the point in the project at which the control is applied. Various control points can be identified at the outset of the project. These will often coincide with the end of stages or modules.

PLANNING

Project plans are raised at the outset of a project and refined as the project proceeds. A high-level plan for the entire project will be supplemented by more detailed plans for each module. There are several types of plan that may be produced, including:

- technical plans, which concentrate on activities and products;
- resource plans, which concentrate on resources and their costs;
- quality plans, which define how quality will be maintained throughout (these are generally incorporated into the technical and resource plans).

Some of the project planning activity is undertaken formally within SSADM stages:

- Step 110 — an Activity Network and Activity Descriptions are produced, together with a Product Breakdown Structure, Product Descriptions and a Product Flow Diagram.
- Step 420 — plans for the development of the system are produced as part of Technical System Options.

All other project planning is undertaken by the managers of the project as the project proceeds.

ESTIMATING

Estimating the time and resources required to complete each SSADM module needs to be based upon experience of the use of SSADM on previous projects. This gives an approximate idea of resources required to complete each of the SSADM products. It is possible to devise 'rules of thumb' based upon the interrelationships between products, but these rules need to be supplemented by experience and common sense to produce realistic plans.

Estimates are influenced by a wide variety of factors, for example:

- size of project;
- size of team;
- complexity of the application area;
- knowledge of the application;
- experience and skills of development staff;
- user involvement;
- previous studies such as strategy or feasibility studies.

Some of these factors are more easily determined than others. In some cases, the factors do not have a linear effect on the estimates. For example, estimated time for completion of tasks should decrease as the size of the team increases; but as the development team increases in size, more time will be required to coordinate and control the work, and this will have a detrimental effect on the estimates.

Estimating needs to be done iteratively as knowledge of the environment and influential factors increases throughout the project.

Quality assurance

Quality assurance is the process by which the acceptable level of quality for each product is defined and tested at defined points in the project. A strategy for quality assurance should be agreed at the outset of the project and planning should be done to ensure that this strategy is applied correctly.

The control of the quality of SSADM products requires a series of reviews during the life of a project. These reviews examine completed products at specific points in the development lifecycle of the project to ensure their adequacy to act as a basis for subsequent activities.

The checks applied at the reviews will concentrate on three aspects:

* Technical correctness — does the product adhere to the rules of SSADM?
* Business correctness — does the product correctly reflect the users' perception of what it represents?
* Fitness for purpose — does the product meet its objectives, either as a deliverable from a module or as the basis for subsequent products?

Reviews may be either formal or informal. A formal review is normally held at the end of each stage and possibly at intermediate points where important products are completed. A formal review requires the products to be formally 'signed off' by the reviewers. Informal reviews often precede formal reviews at an earlier point in the development of a product, or they are used to review intermediate products that do not form part of the product set output from a module.

The reviewers are normally taken from both the project assurance team and the project as a whole. It is important to include representatives from all relevant areas in order that all aspects of quality are checked. The reviewers should therefore consist of:

* technical reviewers to ensure correct usage of SSADM;
* user reviewers to ensure the system meets their requirements;
* representatives of the business to ensure conformance to organizational objectives.

Risk assessment and management

Risk assessment and risk management are related activities which help to determine levels of risks to assets and reduce the vulnerability of those assets to the identified risks.

The levels of risk depend upon the threats identified and the corresponding vulnerability to those threats. The type of countermeasure selected to reduce vulnerability to the risks depends upon the value of the assets that require protection.

The CCTA Risk Analysis and Management Methodology (CRAMM) consists of three stages:

- Stage 1 asks if there is a need for security above a basic level of good practice by identifying physical and data assets and placing a value on them.
- Stage 2 asks what is the security need and where is it needed by identifying threats and vulnerabilities to those threats and assessing risks.
- Stage 3 asks how to meet the needs for security identified in the previous stage by identifying and selecting countermeasures.

Risk assessment and management techniques help to determine what level of security is required for a system. Security can be applied to both the hardware and software of a system. The levels of security required can affect the cost of the system and the technical architecture adopted. This has an impact on the options for a project, especially the Technical System Options.

Capacity planning

Capacity planning is part of the overall discipline of managing the capacity of computer systems. It involves ensuring that sufficient computing capacity exists to meet future demands and that predicted service level requirements can be met in an optimum way.

Capacity planning can help to size the required system during the formulation of options within SSADM. If the right information is made available from within SSADM, capacity planning techniques can be used to:

- check that the current capacity of a computer is sufficient to meet the required service levels and check that any existing service level agreements are not violated;
- estimate the required capacity of a new computer system in terms of both the current and future requirements;
- estimate how much it will cost to meet the required service levels.

The estimation of capacity needs to be based on proposed architectures, so there needs to be some element of Physical Design within the options for capacity planning to be of use.

The inputs required to capacity planning are:

- Volumetrics of both data and processing together with an estimate of the complexity of processing required. This needs to take into account peaks and troughs, concurrency, and the effect of the usage of different routes of processing.
- Service level requirements from the Requirements Catalogue and Function Definitions to give an indication of the users' requirements for how the functionality and data must be delivered to them.
- Proposed Technical Environment Descriptions to describe the architecture of the new system.
- Descriptions of each option from Business System Options.

Capacity planning techniques are used to estimate the size of computer that is required to support each of the options. Where a new application is to be supported by an existing computer, capacity planning techniques help to check that there is sufficient capacity for the new application without degrading the existing applications.

The results of capacity planning are used as an input to the Cost/Benefit Analysis and in assessing each of the options.

Testing

Testing is carried out both by the developers of a system and by the users to ensure that the system meets its objectives and is operationally sound. The testing carried out by developers generally takes place at three levels:

- Unit testing is carried out on the individual components of the system in isolation. Unit testing is usually performed on individual components as soon as they are produced.
- Integration testing is carried out after unit testing has demonstrated that each of the components of the system meets its objectives. The purpose of integration testing is to ensure the components work together as they should.
- System testing, by contrast, normally treats the system as a 'black box' and investigates its behaviour without concern for individual component and internal interfaces.

The testing carried out by the users is acceptance testing. Acceptance testing is carried out to ensure the system as a whole, including each unit and interfaces between units, meets the users' requirements.

The main areas of activity associated with any form of testing are:

- development of test criteria;
- production of test specifications;
- conduct of the test.

Test criteria describe the aspects of the system that are subject to testing and define what constitutes a 'pass' or a 'fail'. The criteria are based on what the user requires of the system and so may be derived from the SSADM documentation produced during the Requirements Specification Module and the Logical System Specification Module. As such, integration testing is normally related to the Logical System Specification, while system testing is related to the Requirements Specification.

Two of the main inputs to test criteria are the Requirements Catalogue and Function Definitions. The service level requirements define a 'target value' and 'acceptable range' for a number of aspects of the system. These can readily be converted into test criteria.

Test specifications are based on the test criteria but are specific to components of the implemented system. Therefore, both the production of test specifications and the conduct of the various tests are carried out in a subsequent part of the development lifecycle to that covering SSADM.

Training

There are two aspects of training that have relevance to SSADM:

- the training required to undertake the project;
- the specification and design of training for the users of the new system.

Training is required for all participants in the project if they are new to the techniques and disciplines being used. The types of training that may be considered are:

- general systems analysis and design skills for analysts and designers;
- SSADM analysis and design for analysts and designers;
- appreciation of SSADM for the users who will be involved;

- quality control of SSADM documentation for the project assurance team;
- project management and configuration management techniques for use with SSADM for the managers;
- capacity planning techniques for use in options;
- risk assessment and management techniques for use in options;
- implementation-specific training for physical design.

The needs for such training should be identified as early as possible to ensure that the right skills are available when they are needed. The training should be planned such that the skills acquired in training are applied in practice soon afterwards to maximize the benefits of the training.

The specification and design of training for the users of the system should be undertaken as early as possible. This will involve the design of training courses and the production of a training manual. User training is often of vital importance in ensuring the most effective use of the system and its final acceptance.

Take-on

'Take-on' is the conversion of data from the current system for use in the new system. The current system data may be held either manually or on a current computer system. It is possible that a decision is made not to use any of the data from the current system, but in the majority of cases a significant proportion of the current data is required by the new system.

The following area may be considered:

- What proportion of the data should be taken on. In some cases, not all the data in the current system is suitable for use in the new system.
- Description of the method of conversion, both in terms of the physical method of transfer between hardware and in terms of any manipulation of the data that may be required to fit into the new data structure.
- Hardware requirements for transfer if it is to be an automatic transfer.
- Specification of validation of the data to ensure error-free data is taken on.
- Resource estimates both for the design of the facilities for take-on and for the actual take-on process itself. It is often possible to underestimate the effort involved in data conversion, especially if the data is to be rekeyed.
- Strategy for cut-over from the old system to the new. If the cut-over needs to be completed quickly, this will have an impact on the method and planning of the take-on.

Take-on should be considered as part of the overall application. The specification and design of this area can be started in parallel with the Requirements Specification and Logical System Specification modules.

(Technical) authoring

There are a number of stages in SSADM that require the production of written reports for evaluation by the users or project board. Also, the documentation produced for use with the new system needs to be written clearly and informatively. The skill of technical authoring is required to ensure effective communication between developers and users. This is especially important if the users are more able to understand written descriptions than diagrammatic techniques.

Two of the major authored outputs of a project are user manuals and operating instructions. The purpose of the user manuals is to instruct the users in the topics required for the successful operation of the system. The manuals should contain a description of the facilities available, constraints imposed, and actions to be taken in the event of errors being encountered. The language used should be readily understood by the users and the layout should be readily accessible. The user manuals may be initiated after the Specification of Requirements is completed and the functions defined.

Operating instructions are used as a reference by the staff responsible for the operation and administration of the new system. The instructions should contain a description of the purpose of the system, the work to be done in its routine operation, action to be taken in the event of failure, and archive procedures. The operating instructions may be initiated after the Technical System Options have defined the technical architecture of the system and confirmed the requirements in the Requirements Catalogue.

Standards

An organization may have a number of different types of standard that have an impact on the use of SSADM:

- Human–computer interface standards in the form of a style guide.
- Project management standards, which define how a project is controlled and quality assured.
- Development standards defining hardware and development software to be used for the new system.
- Documentation standards defining the format and content of reports and manuals.
- Programming and physical design standards that affect the strategy adopted in the Physical Design module.

Any such standards should be taken into account when planning the project. In some cases, standards may conflict with the SSADM steps, stages, and deliverables. In this case, the conflicts should be identified as soon as possible so that they can be resolved.

Information highway

The information highway is the means by which SSADM passes information to and receives information from the various project procedures. This is represented as being in parallel with all the SSADM stages with flows of information to and from particular steps or the entire stage. Figure 14.2 represents this concept.

The information highway also passes information between modules. In SSADM, there is no direct interface between modules. Each module produces a predefined set of outputs which are passed to the information highway. The products are then passed from the information highway to the subsequent module(s) where they are used.

The specific interfaces to and from the information highway are represented in Table 14.3.

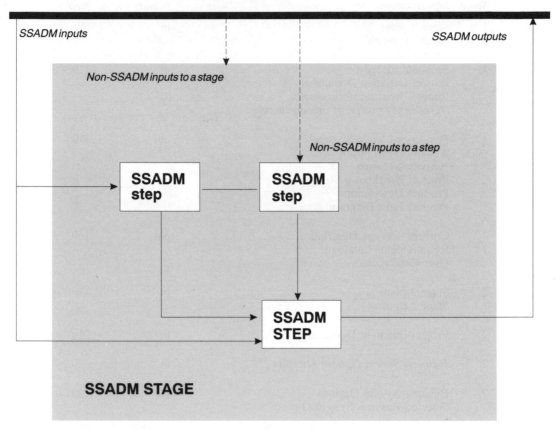

Figure 14.2 The information highway

14.6 Comparison with Version 3

There is no equivalent concept in Version 3 that corresponds to project procedures. However, project procedures do relate to a number of components that are part of Version 3:

- Project management elements are embedded in some of the Version 3 step descriptions, especially planning.
- Quality assurance techniques were included in the final step of most stages in Version 3.
- Test plans are included as a product of Physical Design in Version 3.
- Authoring is involved in the production of user manuals and operating instructions in Physical Design in Version 3.

Table 14.3 Interfaces with the information highway

Stage	Inputs/outputs	In/out	Step
1	Stage 1 plans *Stage 1 controls*	in	all
1	Feasibility Report Project Initiation Document Reports from previous studies *Agreement to scope of investigation*	in	110
1	*Project and analysis scope* Activity Descriptions Activity Network Product Breakdown Structure Product Descriptions Product Flow Diagram	out	110
1	Current Services Description Requirements Catalogue User Catalogue	out	160
2	Stage 2 plans *Stage 2 controls*	in	all
2	Project Initiation Document	in	210
2	*Business System Options selection*	in	220
2	Business System Options Selected Business System Option	out	220
3	Stage 3 plans Data Catalogue *Stage 3 controls*	in	all
3	Logical Data Flow Model Logical Data Store/Entity Cross-reference User Catalogue	in	310
3	Requirements Catalogue Selected Business System Options	in	310 320
3	Current environment Logical Data Model	in	320
3	Installation Style Guide Prototyping Scope	in	350
3	Requirements Specification	out	380
3	Command Structures Menu Structures Prototyping Report	out	350

Stage	Inputs/outputs	In/out	Step
4	Stage 4 plans *Stage 4 controls*	in	all
4	*Evaluated capacity planning information* Project Initiation Document Requirements Specification Selected Business System Options	in	410
4	*Evaluated capacity planning information* Installation Style Guide *Technical System Options selection*	in	420
4	Capacity planning input	out	410
4	Technical System Options Application Style Guide Capacity planning input Technical Environment Description	out	420
5	Stage 5 plans *Stage 5 controls*	in	all
5	Function Definitions I/O Structures Requirements Catalogue Style Guide User Role/Function Matrix	in	510
5	Effect Correspondence Diagrams Entity Life Histories Function Definitions I/O Structures Required System Logical Data Model Style Guide	in	520
5	Enquiry Access Paths Function Definitions I/O Structures Relationship Descriptions Required System Logical Data Model Style Guide	in	530
5	Effect Correspondence Diagrams Elementary Process Descriptions Enquiry Access Paths Function Definitions I/O Structures Required System Logical Data Model User Role/Function Matrix	in	540
5	Logical Design	out	540

Stage	Inputs/outputs	In/out	Step
6	Stage 6 plans *Stage 6 controls*	in	all
6	Technical Environment Description *Installation development standards*	in	610
6	Effect Correspondence Diagrams Enquiry Access Paths Function Definitions Required System Logical Data Model	in	620
6	Effect Correspondence Diagrams Enquiry Access Paths Function Definitions Enquiry Process Models Requirements Catalogue Update Process Models	in	640
6	Logical Design	in	630 650
6	Required System Logical Data Model	in	660 670
6	Application Development Standards	out	610
6	Physical Design	out	670

in = input to step or stage from the information highway.
out = output from step or stage to the information highway.
Items in italics are non-SSADM inputs and outputs.

Appendix A. Version 4 Structural Model

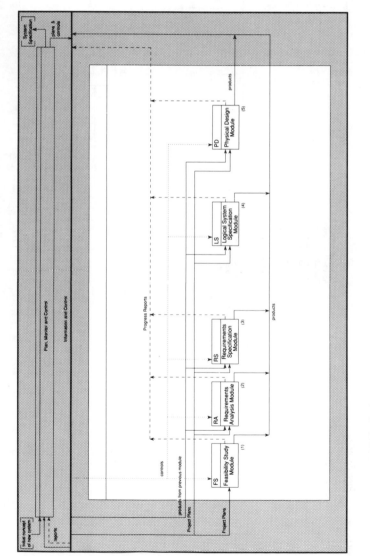

Figure A.1 SSADM Life Cycle

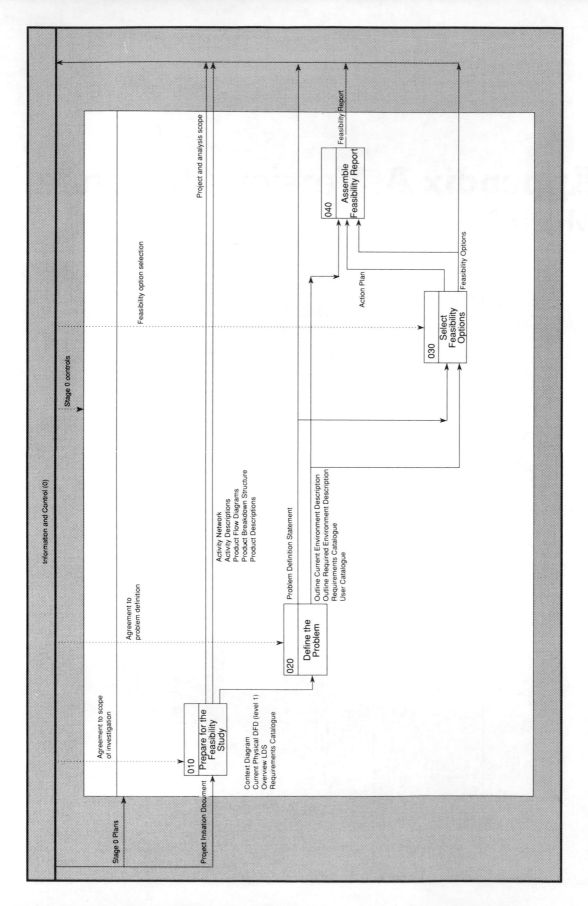

Figure A.2 Feasibility Study Module: Stage 0 Feasibility

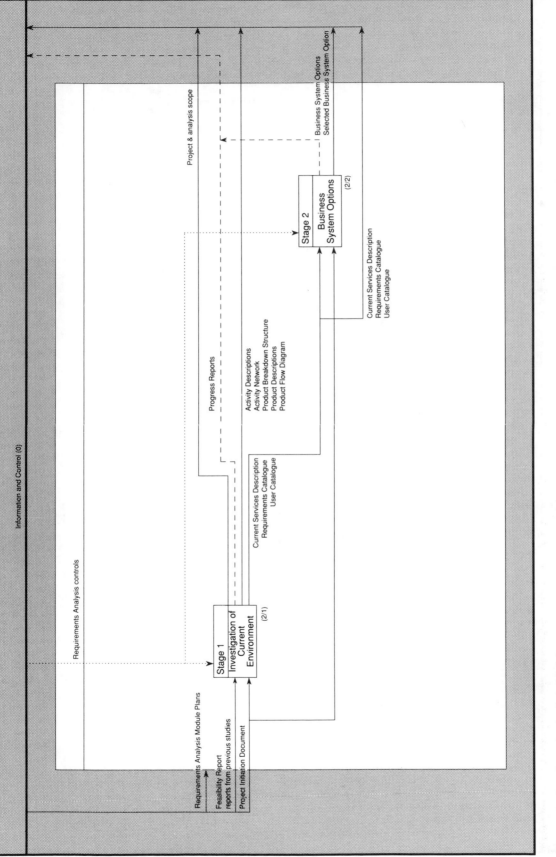

Figure A.3 Requirements Analysis Module

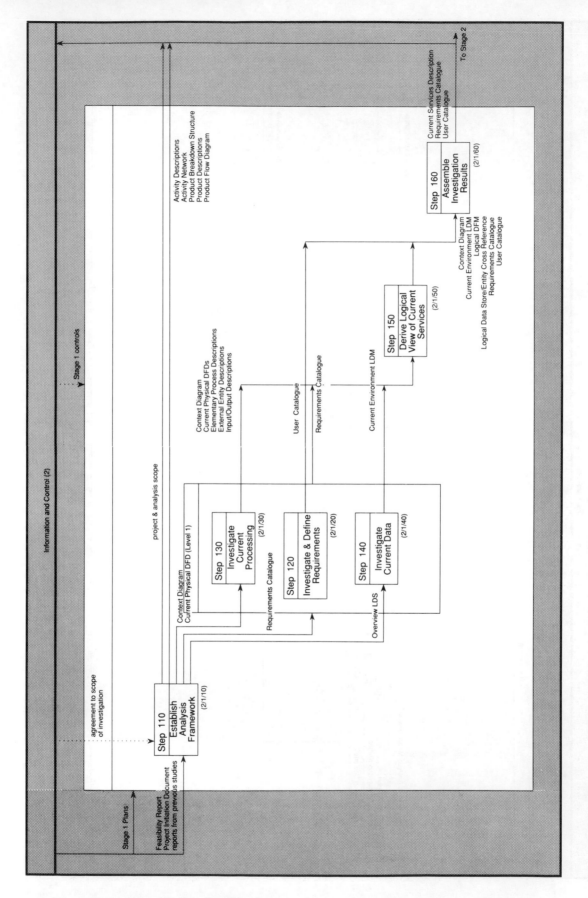

Figure A.4 Requirements Analysis Module Stage 1: Investigation of Current Environment

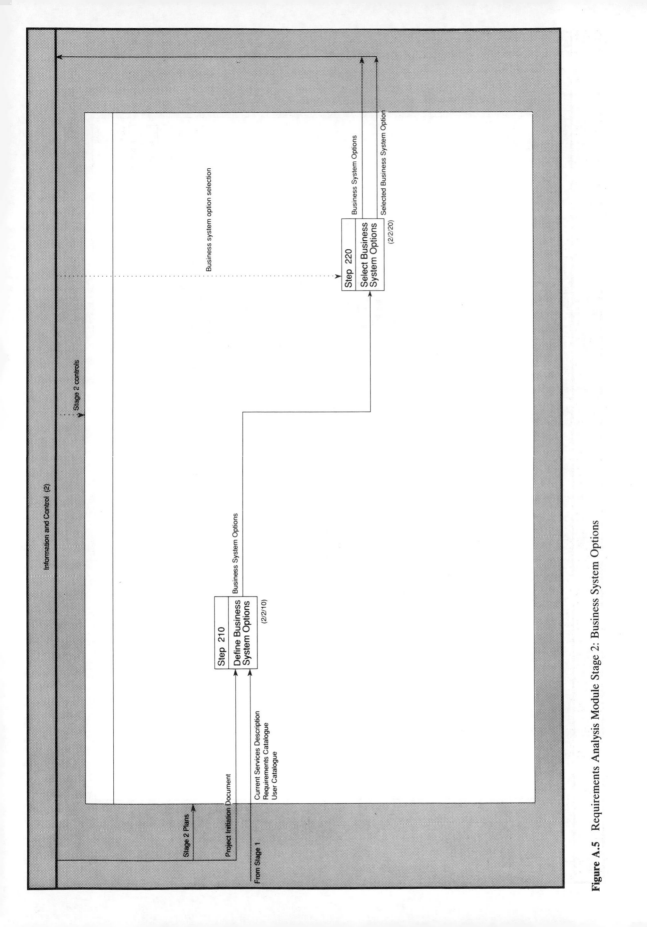

Figure A.5 Requirements Analysis Module Stage 2: Business System Options

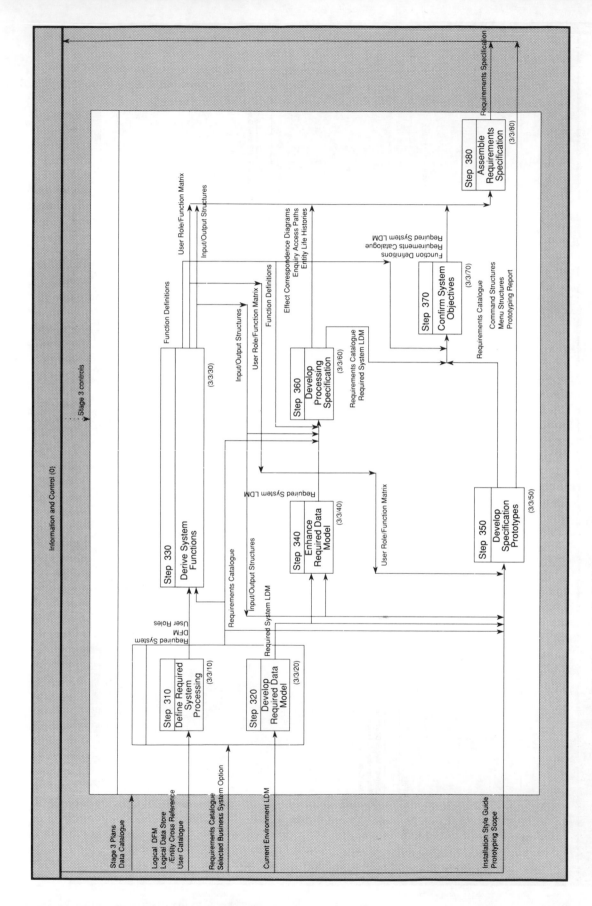

Figure A.6 Requirements Specification Module Stage 3: Definition of requirements

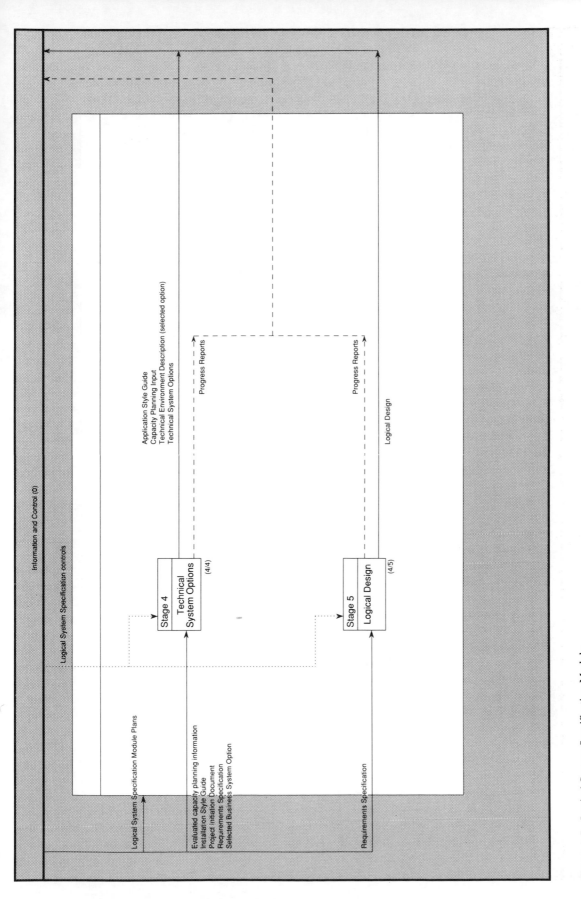

Figure A.7 Logical System Specification Module

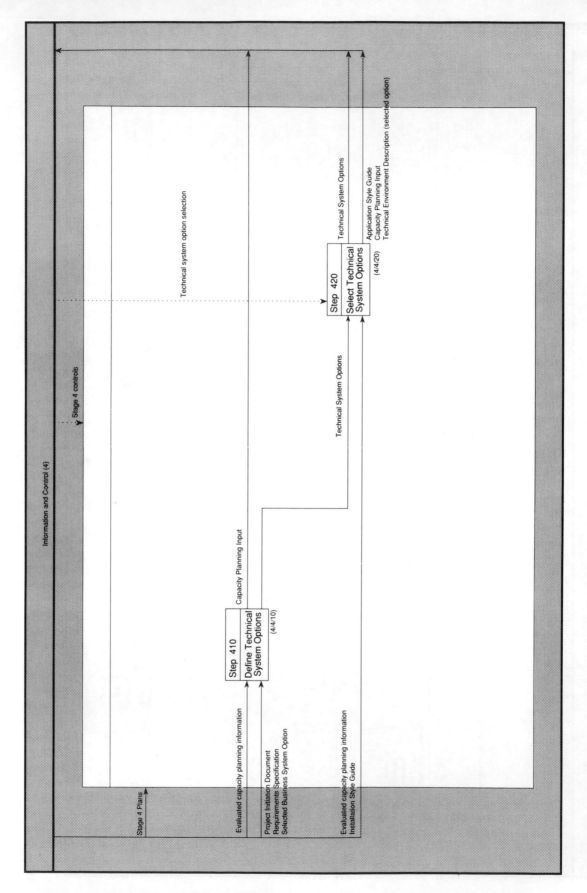

Figure A.8 Logical System Specification Module Stage 4: Technical System Options

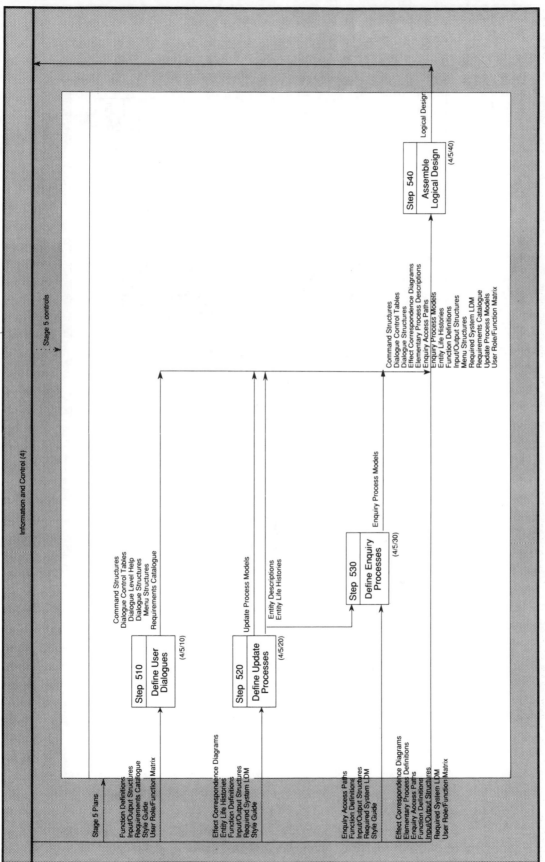

Figure A.9 Logical System Specification Module Stage 5: Logical Design

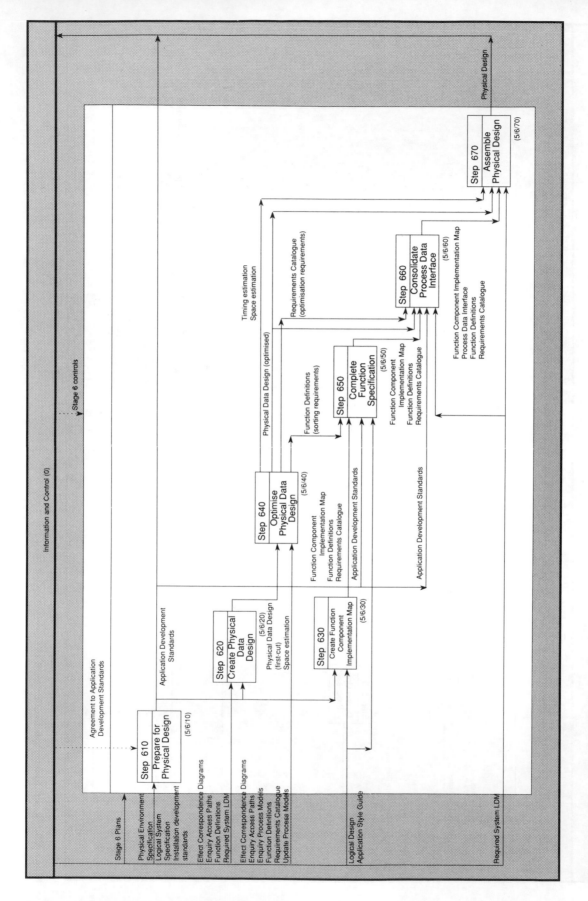

Figure A.10 Physical Design Module Stage 6: Physical Design

Appendix B. Product Breakdown Structure

This appendix gives a diagrammatic view of the complete Product Breakdown Structure for the SSADM products.

There are five diagrams. The first of these shows how the outputs from each of the modules fits into the overall Product Breakdown Structure for the project products. There then follows one diagram for each module's output with the exception of the Feasibility Report, which is not broken down further.

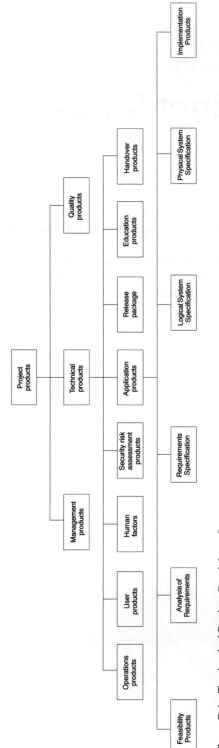

Figure B.1 Top level of Product Breakdown Structure

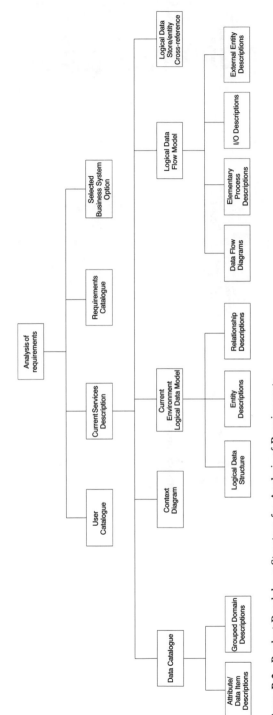

Figure B.2 Product Breakdown Structure for Analysis of Requirements

207

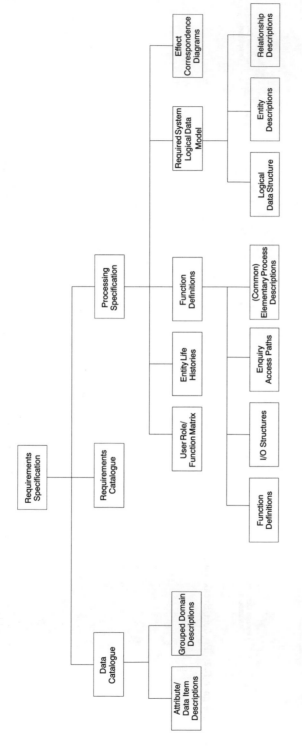

Figure B.3 Product Breakdown Structure for Requirements Specification

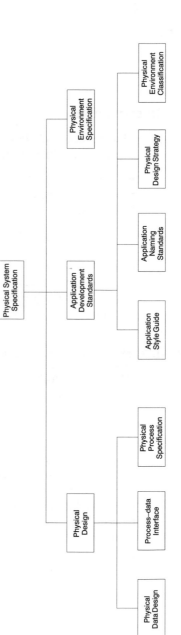

Figure B.4 Product Breakdown Structure for Logical System Specification

Figure B.5 Product Breakdown Structure for Physical System Specification

Appendix C. Products Covered by the Dictionary

C.1 Introduction

The SSADM Dictionary — a new concept introduced into Version 4 manuals — is divided into three sections:

- Product Breakdown Structure;
- Product Descriptions;
- Glossary.

The first of these is dealt with in this appendix and also within each of the chapters dealing with the techniques. The Product Descriptions give a definition, description, and, if necessary, a form for each of the products within SSADM Version 4. The Glossary gives a definition for the terms used within SSADM Version 4.

Within this appendix is a list of the products covered by the second section of the Dictionary.

C.2 List of Product Descriptions Covered by the Dictionary

Activity Descriptions
Activity Network
Analysis of Requirements
Application Development Standards
Application Naming Standards
Application Style Guide
Attribute/Data Item Descriptions
Business System Options
Capacity Planning Input
Command Structure
Context Diagram
Cost/Benefit Analysis
Current Services Description
Data Catalogue
Data Flow Diagram — Level 1
Data Flow Diagrams — Lower Level
Data Flow Model

DBMS Data Storage Classification
DBMS Performance Classification
Dialogue Control Table
Dialogue Element Descriptions
Dialogue Level Help
Dialogue Structure
Dialogues
Document Flow Diagram
Effect Correspondence Diagram
Elementary Process Descriptions
Enquiry Access Path
Enquiry Process Models
Entity Descriptions
Entity Life Histories
Event/Entity Matrix
External Entity Descriptions
Feasibility Options
Feasibility Report
Function Component Implementation Map
Function Definition(s)
Generic Blank Form
Generic Matrix Form
Grouped Domain Descriptions
Impact Analysis
Installation Style Guide
I/O Descriptions
I/O Structure
I/O Structure Description
I/O Structure Diagram
I/O Structures (for all functions)
Logical Data Model
Logical Data Store/Entity Cross-reference
Logical Data Structure
Logical Design
Logical/Physical Data Store Cross-reference
Logical Process Model
Logical System Specification
Menu Structure
Outline Development Plan
Physical Data Design
Physical Design
Physical Design Strategy
Physical Environment Classification
Physical Environment Specification
Physical Process Specification
Physical System Specification
Plans

Process–Data Interface
Process/Entity Matrix
Processing Specification
Processing System Classification
Product Breakdown Structure
Product Descriptions
Product Flow Diagram
Progress Report
Project Initiation Document
Prototype Demonstration Objective Document
Prototype Pathway
Prototype Result Log
Prototyping Report
Prototyping Scope
Relational Data Analysis Working Paper
Relationship Descriptions
Report Format
Requirements Catalogue
Requirements Specification
Resource Flow Diagram
Screen Format
Selected Business System Option
Selected Technical System Option
SSADM Structure Diagram
System Description
Take-On Requirements Description
Technical Environment Description
Technical System Options
Testing Outline
Training Requirements Description
Update Process Models
User Catalogue
User Manual Requirements Description
User Role/Function Matrix
User Roles

Appendix D. Structure Conventions for SSADM Diagrams

D.1 Introduction

A number of SSADM techniques are based upon the same set of structuring rules. These basic rules are modified and added to by each of the techniques but the basic conventions are the same wherever they are used. The chapters of the book describe each of the techniques and their application. This appendix lays down the basic syntax that is common to all techniques and describes briefly the representation used for each technique and any variations or extensions.

D.2 Structuring conventions

The structuring conventions are based upon the Jackson structuring technique and have the following structuring components:

- sequence;
- selection;
- iteration.

These components can be combined in a variety of ways to form simple or complex structures representing life histories, structures of input and output data, dialogue interactions and process structures.

Sequence

A sequence is represented by a series of boxes reading from left to right as shown in Fig. D.1. The box labelled A is always the first to occur, followed by B, which in turn is followed by C and D. This is the only possible sequence. There is no other dependency between the boxes, e.g. A does not trigger B, and there is no implied time interval between the two.

Selection

A selection defines a number of alternatives at a particular point in the structure. A selection is represented by a set of boxes with circles in the top right corners as shown in Fig. D.2. Only one of the boxes may be selected at this point in the structure.

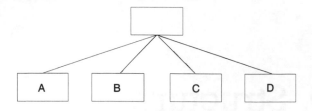

Figure D.1 Sequence

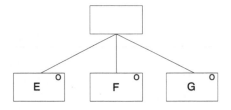

Figure D.2 Selection

If there is a need to show that none of the alternatives needs to be selected, a 'null box' may be added as shown in Fig. D.3. The null box does not represent anything being done. It is simply a notational device to indicate that nothing may happen at that point and the sequence will resume at the following box.

Iteration

An iteration represents the repetition of a box from zero to many times at a particular point in the structure. A restriction on the iteration is that each occurrence of the iteration must be complete before the next begins. An iteration is represented by an asterisk in the top right corner of a box as shown in Fig. D.4.

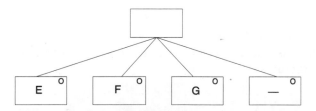

Figure D.3 Selection with null box

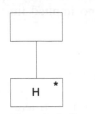

Figure D.4 Iteration

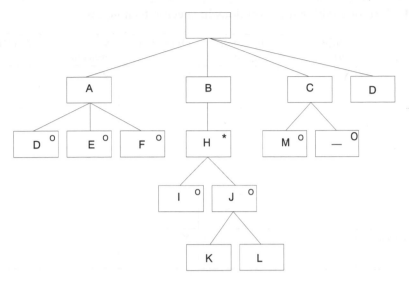

Figure D.5 Combination of structure elements

Combination of the components

The basic components of sequence, selection and iteration can be combined into a structure as demonstrated in Fig. D.5. There are several things to note about such structures:

- Only one type of structure convention can be used at the same level.
- The boxes with other boxes below them are 'nodes' and are only present to aid the structure.
- The boxes at the bottom of the structure are 'elementary' and represent whatever is being modelled.

D.3 Technique-specific points

The use of the standard structuring conventions is tailored to fit each circumstance within which it is applied. Each technique has a slightly different application of the conventions. This leads to the addition of extra conventions in some techniques and different naming standards. Also, the use of 'hard' and 'soft' boxes is technique-specific.

Additional conventions

There are a number of additional conventions that are used in specific techniques in addition to the basic conventions described above. The additional conventions add greater flexibility or extra information to the basic notation for individual techniques.

The additional conventions used are:

- parallel structures;
- quits and resumes;
- operations;
- state indicators;
- logical groupings of dialogue elements.

Table D.1 Use of additional conventions in specific techniques

	Parallel structure	Quits and resumes	Operations	State indicators	LGDEs
Entity Life Histories	Y	Y	Y	Y	
I/O Structure					
Dialogue Structure					Y
Update Process Model			Y		
Enquiry Process Model			Y		

Table D.1 shows which of the additional conventions are used in which techniques. The usage of these additional conventions is described in the relevant chapters.

Please note that although Effect Correspondence Diagrams and Enquiry Access Paths use the selection and iteration components of the structuring notation, they are not included here as they do not conform to the sequence rules and are not built in a hierarchical way as the other diagrams are.

Naming and type of 'elementary' boxes

The bottom boxes on a structure (excluding the operation boxes) are the building blocks of the structure and have different meanings depending upon their context. Also, the shape of the elementary boxes is different depending upon the context. Table D.2 summarizes the naming and symbol conventions for the different techniques.

Table D.2 Naming and type of elementary boxes

	Elementary box name	Soft/hard
Entity Life Histories	Effect	H
I/O Structure	I/O Structure element	S
Dialogue Structure	Dialogue element	S
Update Process Model	Not named	H
Enquiry Process Model	Not named	H

Appendix E. Version 3 and Version 4 Technique Mapping

Version 3	Version 4
Logical Data Structuring	Logical Data Modelling
Composite Logical Data Design	
Data Flow Diagramming	Data Flow Modelling
Problems and Requirements List	Requirements Definition
Entity Life Histories (subset)	Entity/Event Modelling
—	Effect Correspondence Diagramming
Entity Life History Analysis	Entity Life History Analysis
—	Enquiry Access Paths
Function Definition	Function Definition
Relational Data Analysis	Relational Data Analysis
—	Specification Prototyping
Business System Options	Business System Options
Technical Options	Technical System Options
Logical Dialogue Design	Dialogue Design
Logical Process Design	Logical Database Process Design
Program Specification	Physical Process Design
File/Database Design	Physical Data Design

Appendix F. Description of Opera Booking System

F.1 Introduction

The Opera Booking System is designed to assist the operation of the box office of a fictional opera house. The only areas covered are to do with the allocation and issuing of tickets for opera performances to customers. Other areas of the operation of the opera house that interface with the Opera Booking System are:

- the engaging and payment of performers;
- the setting up of production details;
- the general finances of the opera house;
- the seating plan for the opera house.

F.2 Users

There are four main categories of users of the new system. The *Opera House Manager* is the man at the top. It is his responsibility to ensure that the opera house runs efficiently as a whole. He is ultimately responsible for organizing the programme of productions each season, the hiring/firing of singers, producers, designers and all the other people involved in the staging of operas, and determining the pricing policy.

Beneath the Opera House Manager there are a number of *House Day Managers* to whom the Opera House Manager delegates various responsibilities to do with the day-to-day running of the opera house. In addition, one of the House Day Managers takes responsibility for managing the Box Office.

Within the Box Office itself there are two different types of clerk. *The Booking Office Clerks* sit at the windows of the box office and deal with the members of the public who are either making enquiries about performances or are booking tickets for specific performances. This can either be in person or by telephone. At the back of the box office sits the *Postal Office Clerk* whose responsibility is to process postal applications and to send out tickets once the seats have been allocated.

F.3 Information Held

The information held is needed to support the processing of orders, the allocation of seats to customers and the production of tickets. In addition it must also assist enquiries from members of the public and the opera house management.

The system needs to keep information from orders sent in by post so that any problems that might arise can be traced back to their original order. Each order can consist of a number of lines. Each line will refer to only one performance but may request a number of seats.

In order to allocate seats and print tickets the system needs to contain details of all seats in the theatre that may be booked for each performance. This information is divided into two separate parts:

- details of each physical seat in the theatre;
- details specific to the use of that seat at a performance.

In order to allocate seats to orders, the system simply links the details of a specific seat to a specific order line. The individual order lines will specify only a particular part of the theatre, rather than specific seats. Therefore the system must hold details of which seats are in which area.

To satisfy enquiries the system holds minimal information about the productions for a particular season and the artists appearing in them. A production is a run of a particular opera in the relevant season. Artists are engaged to sing in all performances of a production. A production has a number of performances during the season. Potentially the seats available can vary from performance to performance, even for the same production. Although a production must be defined in terms of number of performances it is possible for a performance not to be associated with a production. This occurs with one-off performances, e.g. concerts given by the orchestra.

F.4 Processing Supported

The main areas of processing are the maintenance of standing data (e.g. production, artist, theatre details), enquiries about performance details, and the main business of the system, which is the requesting and provision of tickets.

Tickets can be obtained in one of three ways:

- sending an order by post with payment;
- going to the box office in person and paying the Booking Office Clerk;
- phoning the box office and giving a credit card number over the phone.

The processing of orders is more complicated than the other two because at the time the order is received it is not known whether the order can be satisfied, and an order can cover a number of performances. The buying of tickets from the box office and the phoning of orders results in an immediate ticket print and no order details are held on the system except that a seat has now been designated as occupied.

When an order is received by post, it needs to be validated in a number of ways:

- the production details and dates of performances are correct according to the latest information held in the box office;
- the area of the theatre specified is available for booking;
- either a cheque is enclosed or credit card details are provided, together with a signature.

The next step is to try to satisfy the order line by line. Each line specifies a particular performance and can ask for a number of seats. If seats are available, these are allocated to the order line and details are held temporarily, ready for a ticket print for the whole order. If an order line cannot be satisfied, the next line is processed and the Postal Booking Clerk notes the details to send to the customer with any tickets from the order. The order contains the total price as calculated by the customer. If this differs from the total calculated by the system, this is highlighted to the customer. Once all order lines for an order have been processed, the system prints a batch of tickets, which are then sent to the customer. If a credit card number is supplied on the order and the order total is in excess of £100, the tickets are held pending the results of a credit card check carried out by the Postal Booking Clerk via the telephone.

Tickets can be exchanged for different seats or different performances of the same production up until the time of the first performance listed on the order. Once this takes place, no further exchanges are allowed for the whole order.

Details of the customer taken from the order are held on the system in case any last-minute changes to performance details occur which need to be notified to the customer, e.g. if a performance is cancelled. The customer can notify the box office of any changes to their personal details over the lifetime of the order. Any changes notified after the last performance on the order has taken place are ignored as they are no longer of relevance.

Index